D0408448

Spaceflight in the Shuttle Era and Beyond

Spaceflight in the Shuttle Era and Beyond

Redefining Humanity's Purpose in Space

Valerie Neal

Yale UNIVERSITY PRESS | NEW HAVEN AND LONDON

Yale University Press, in association with the Smithsonian National Air and Space Museum.

Yale University Press books may be purchased in quantity for educational, business, or promotional use. For information, please e-mail sales.press@yale.edu (U.S. office) or sales@yaleup.co.uk (U.K. office).

Set in Adobe Garamond and The Sans types by Newgen North America.
Printed in the United States of America.

Library of Congress Control Number: 2016957558
ISBN 978-0-300-20651-7 (hardcover : alk. paper)

A catalogue record for this book is available from the British Library.

This paper meets the requirements of ANSI/NISO Z39.48–1992 (Permanence of Paper).

10 9 8 7 6 5 4 3 2 1

To my son
Bryan Guido Hassin

Contents

Preface

A book on spaceflight began to form in my mind when the space shuttle and my career launched simultaneously; my professional life spans the shuttle era. For thirty-plus years, I have worked in various roles on the periphery of the ambitious endeavor of human spaceflight. What has that meant? The question is both biographical and cultural. This book is my effort to discern the cultural meaning of human spaceflight—its formation and transformations—in this era.

I completed graduate school in interdisciplinary American studies in the 1970s under the influence of the "myth and symbol" tradition of intellectual and cultural history. This approach to American culture through the humanities analyzed the history of ideas and their synthesis in literature and the arts to illuminate broad themes in American experience and thought. From the social sciences and history of science came other intellectually fertile concepts for understanding how meaning is created, understood, codified, and modified; "paradigm shifts," "social construction," "framing," and "imaginaries" entered the scholarly lexicon. Innovative scholarship and analytical trends in humanities and social sciences research continue to invigorate the study of American culture. Conceptual tools and terminology keep changing, but understanding what things mean and how meaning shifts remains a priority.

Educated and predisposed to seek connections between ideas and images, and to read icons as their incarnation, I offer this book in the ever renewing and expanding tradition of culture studies. My focus of inquiry here is a particular American enterprise: human spaceflight in the shuttle era and beyond. In search of its meaning, I explore where answers may be discovered by examining its texts and images and icons, the motives of people and institutions that shaped and spread them, and representations of spaceflight in the broader community. I study its science, technology, and rhetoric. I trace its ebbs and flows and persistence. I approach spaceflight as a cultural text and iconography to be probed and revealed.

Emerging from academics, in the 1980s I worked as a writer-editor under contracts with NASA to support a variety of shuttle missions and science programs. I spent much of my time and energy with mission managers and scientists, jointly creating publications to explain human spaceflight and scientific activities to the public. My job was essentially translation, crafting language and imagery to communicate from a specialized technical world to the world at large. Since I joined the Smithsonian National Air and Space Museum as a space historian and curator in 1989, I have continued such communication with the public through various channels, notably exhibitions and programs about spaceflight. Spending three decades working in space history as it happens is certainly a spur to analysis and reflection.

And so, this book has its origins in my professional experiences where personal narrative intersects with a compelling cultural narrative. Conversant in academic traditions and in spaceflight, I offer here an interdisciplinary perspective on an endeavor that ranges beyond technology, operations, and policy. Human spaceflight means more than that.

Acknowledgments

Throughout this project and others, I have had the benefit of stimulating colleagues at the Smithsonian National Air and Space Museum whose knowledge and interests are an inspiration. First among those whom I credit for influences large and small in our work together, and especially for encouragement of this book, is Michael J. Neufeld, a meticulous researcher, gifted writer, and incisive editor who models the discipline of scholarship in history. Another is Roger D. Launius, a prolific space historian whose range of inquiry is boundless and who writes as easily as breathing. Both have in various ways encouraged my progress on this book, not least by their probing questions and critiques. John R. Dailey, the director of the museum, has also encouraged this effort by expressing his appreciation for the myriad other projects I managed to accomplish at the same time yet nudging this in priority.

The women of the museum—the few female historians and the many women in other roles, past and present—have offered encouragement in more personal ways. I am grateful for their many gestures of interest and support. These friendly cheerleaders have listened, offered advice, made me laugh, and generously lifted my spirit. I especially appreciate our former publications chief, Patricia J. Graboske, for connecting me with Yale University Press.

Peer reviewers, both known and unknown to me, made fine suggestions that challenged me to think more broadly and deeply about certain aspects of my analysis and steered me to sources I had not yet examined. I greatly appreciate insightful reviews by William P. Barry, Linda Billings, Amy E. Foster, James R. Hansen, Matthew H. Hersch, John M. Logsdon, Howard E. McCurdy, and Ronald L. Pitcock at various stages from initial proposal to final manuscript, as well as the anonymous reviewers solicited by Yale University Press and its editorial review board. All contributed to making this a more solid book; they are, of course, blameless for any shortcomings but may claim a share of any merits of my work.

I conducted much of my research in the NASA Historical Reference Collection in Washington, D.C., whose staff is a national treasure. Jane H. Odom, Colin A. Fries, John H. Hargenrader, and Elizabeth M. Suckow have retrieved countless archival files with dispatch, coached me in the use of their indexes and equipment, and helped bring this book to fruition. The staff of NASA's Johnson Space Center History Archives and the JSC Oral History Project has eased access to those rich collections. Whether searching sources online or in its quiet reading rooms, I am always rewarded by the scope and accessibility of the amazing Library of Congress. Staff of the Ronald Reagan and Richard M. Nixon Presidential Libraries responded helpfully to my inquiries and visits. Constance L. "Connie" Moore, senior photo researcher at NASA Headquarters, and Mary J. "Jody" Russell, formerly of the NASA Media Resource Center (photo archives) in Houston, efficiently provided NASA images in proper formats.

The entire National Air and Space Museum Archives and Photography staff helped immensely with illustrations, as did Gregory K. H. Bryant of the registrar's office, Jo Ann Morgan of the Space History Department, and Diana Zarick, the Smithsonian's licensing attorney. Smithsonian behind-the-scenes volunteer and museum docent Brad Marman, a retired public affairs officer, was a great help in researching news coverage and editorial cartoons, and museum librarian Chris Cottrill found elusive publications. Former NASA executive Alan Ladwig graciously gave me his collection of the agency's publicity materials on the shuttle and space station. I also appreciated the assistance of several undergraduate interns on research tasks, especially Lynn Atkin, Mary Bergman, Vickie Lindsey, Claire Pope, and Jordan Wappler.

I have been most fortunate that Joseph Calamia of Yale University Press cultivated this book for publication; his constructive guidance and positive nature kept me on track. He, Samantha Ostrowski, Joyce Ippolito, Susan Laity, Nancy Ovedovitz, and Mary Valencia edited and designed an appealing volume that I trust will please readers as much as it pleases me. Freelancer Bob Land proofread the book and prepared the index.

My family and closest friends have always granted me their confidence, love, and patience, especially as I too often kept working when they invited me to join them. With this book now published, I look forward to a more active social life with Bryan Hassin and Katie Barrett, Patrice Neal and family, Janet Neal Fotioo and family, Julia Lee and Sam Wood, David and Marilyn Thomas, Susan and Jerry Nilsson-Weiskott, Ann and Charles Florsheim, Kathryn and Paul Farmer, Rebecca and Jack Stokes, the Hassin sisters, my California cousins, and the larger circle of good souls who grace my life.

I dedicate this book to my son Bryan Guido Hassin, my first and best contribution to the space shuttle era. Born a year before my career began and two years before the first shuttle launch, he grew to manhood knowing about missions, astronauts, space science, and other aspects of his mom's work. As a child, he often drew and dreamed about spaceflight beside me while I worked at the dining room table or office desk. He never resented my research trips that impinged on his young life; he was stoic, even gallant, about the pursuit of knowledge and adventure that lured me away. Together we attended two space shuttle launches, and he joined me to welcome *Discovery* to the National Air and Space Museum. What a great ride we had together in the shuttle era!

Spaceflight in the Shuttle Era and Beyond

Introduction

According to public opinion polls, Americans are rather fickle about space exploration. In open-ended questions—"Do you think the United States should explore space?"—most eagerly say yes. To more focused questions—"Do you think the United States should explore space or tackle [insert any social issue here]?"—many supporters defect. This suggests that people generally do not have a firm commitment to the meaning or value of space exploration, and particularly not to its higher-risk, higher-cost mode: human spaceflight. Yet most do carry around some kind of mental construct—a metaphor, a meme, a cliché—that gives spaceflight meaning in their own intellectual domains.[1]

This book probes the public meaning of human spaceflight during the space shuttle era by examining a variety of constructs or frameworks that shaped and communicated its rationale. Historians, anthropologists, and social scientists for some time have been influenced by a philosophy of knowledge that posits the invention of meaning. That is, the concepts by which humanity lives originate in societies' needs; social practices and their meaning are deliberate creations to serve those needs, and thus may be formed and transformed intentionally over time. Concepts and meaning may be conveyed via narratives, myths, symbols, images, icons, rituals, and traditions, as well as direct discourse. Some meanings endure, some are adapted, and some fail.[2]

Spaceflight is such an invention. It is a malleable concept whose meaning is consciously framed by individuals and groups to fit their circumstances and to influence supporters, opponents, and decision-makers. From President Richard Nixon's 1972 announcement of the decision to launch the shuttle program, to the 2012 deliveries of the retired shuttle orbiters to museums, and into the years beyond, interested parties have defined, doubted, and debated the meaning of human spaceflight. At the extremes, people have argued that it is a grand endeavor to fulfill America's and humanity's destiny, or conversely a lavish misappropriation of resources better directed to resolve pressing social needs. The space shuttle, astronauts, and International Space Station are three icons that bear these meanings, both favorable and critical.

This study started with basic questions about the origins and evolution of meaning. It became an exploration into a variety of source materials and bodies of research where relevant academic disciplines intersect in the study of language, images, thought, and culture. What began as a literary approach—primarily rhetorical analysis of verbal and visual texts—led to suggestive studies in the social sciences that focused on organizations and communications. Analytical tools and methods discovered there offered other lenses to scrutinize the meaning of spaceflight.

To fulfill its educational mission (and to remain popular), the National Aeronautics and Space Administration (NASA) has generated a prodigious amount of information and positive messages about its activities, especially human spaceflight, that is rich source material for this study. Its trove of public and internal documents on the reasons for and benefits of a human presence in space is evidence of the thoughtful, intentional construction of meaning in order to inform and earn support. This body of information potentially influences everyone from the White House and Congress to households and classrooms around the country and, via television and the internet, around the world. The crafting of materials having such tremendous reach is quite deliberate; they issue from the soul of the spaceflight enterprise, from experts working together to decide what to write, drafting, reviewing, word-smithing, and choosing illustrations until the product expresses exactly what they want to convey. Because the massive volume of available NASA materials is overwhelming, I drew the boundary for this book around printed content: documents, speech transcripts, correspondence, publications, photographs, and the like. I included reports that are archived and accessible via the Web but not materials created for that medium or for broadcast; those extensive sources would warrant another book. I did, however, include the IMAX films shot in space in cooperation with NASA as uniquely vivid and deliberate representations of spaceflight. These varied sources rely on verbal and visual rhetoric to promote human spaceflight.[3]

To tap into the external part of the communications loop that responds or "talks back" to NASA's information, I examined two primary sources. One is the news media, also limited to print to permit a more equivalent comparison of responses to NASA's activities, and the other is a body of reports by external advisory and review committees specifically tasked to evaluate concepts put forth by NASA. Both the media and review committees are well briefed by NASA but are responsible for making their own judgments about the meaning of human spaceflight; they likewise craft their messages with great care for public consumption. Their work gives evidence of intended public perceptions of human spaceflight, or at least those of some informed public, and of alternative frames of meaning also meant to persuade. During the shuttle era, these sources of meaning often challenged NASA's. For news coverage and editorial opinions on spaceflight and space policy matters, I relied most on the *New York Times* and *Washington Post* as national organs with broad reach throughout the country and abroad, and thus among the most carefully cultivated by NASA. The media in general, and the *Times* and *Post* as elite media, have significant influence in setting the public agenda and framing issues for public consumption. I did not mine the abundant radio and television coverage of spaceflight in the shuttle era, resources vast enough to prompt another book. However, there can be no doubt of the power of the broadcast media to capture public attention and influence attitudes about spaceflight, just as they do on any other issue.[4]

This study is inspired also by social sciences work in organizational psychology and communications, particularly research that illuminates issue-framing strategies and metaphors deployed to win support and motivate action. Conceptual and relational thinking, traditions, language, and metaphor can be used as tools for constructing intellectual frameworks that hold and communicate meaning. Such research points to the effort that goes into developing a theme for a strategic plan, presidential address, or public relations campaign or into fueling its opposition. Research in linguistics, rhetoric, and cognitive psychology often attends to the uses of symbols and metaphors, with insights that are relevant for discerning the meaning of spaceflight. Also suggestive is scholarship in cognition and memetics that examines memes—word phrases, narratives, symbols, traditions, or behaviors that are imitated and transmitted through the cultural environment. "Space frontier" references, the thumbs-up gesture often used by astronauts, and the depiction of the space shuttle as a truck are examples of cultural memes.[5]

Primarily from anthropology and sociology, with an infusion from philosophy, comes an evocative concept: the imaginary, a noun rather than an adjective. This fairly recent arrival in culture studies has resonance with more traditional terms like belief and myth. An imaginary is the broad common understanding that permeates a society and makes sense of its norms and practices. Some call it a cultural narrative or myth that explains a people's identity and place in the world. Others define imaginary as the widely shared "background knowledge" of why things are as they are, a consensus that gives civic life purpose and coherence, that people implicitly understand without knowing quite how they learned. Imaginaries are historical constructs that evolve with society. They arise from real experience but their factual basis accrues layers of meaning that affect, and are affected by, public attitudes, policies, and actions. Spaceflight is such an imaginary, a matrix of ideas and images that is widely shared and understood but not fully explicable.[6]

The interdisciplinary study of American culture taps into these concepts and more as avenues of inquiry. Language is the primary medium of exchange for cultural identity, narratives, discourse, and imaginaries, but images and icons also signify meaning. Visual culture and visual rhetoric are important channels of communication that represent ideas and values and thus convey meaning. The vibrant field of visual culture studies within the humanities and social sciences that seeks to interpret the meaning of visual "texts" is also relevant to this study of spaceflight.[7]

In the shuttle era, various actors promoted and perceived spaceflight in much different frameworks or imaginaries than in the 1960s. Its justification shifted from a highly optimistic new vision in the 1970s, through struggles and turmoil in the 1980s and 1990s, to uncertainty about an American future in space in the first fifteen or more years of this century. This book addresses the shaping influence of ideas, images, and icons—the verbal and visual rhetorics—of spaceflight in the shuttle era. Each chapter examines a different frame of reference or imaginary, setting it in context and exploring the images and icons that give it substance. The chapters start with a snapshot of a moment in space history that establishes a theme and opens the door to analysis and interpretation.

While my approaches and interpretations are inspired by the varieties of scholarship just mentioned, I have not adhered to a precise methodology from any one of them. My research is not based on quantitative analysis, but on close reading and synthesis. I use such terms as imaginary and discourse somewhat liberally and suggestively, but with care. This is fair use, I think, because as these concepts spread out across scholarly disciplines, they gain or lose nuances according to the norms of that community. For the purposes of this book, it is more useful and satisfying to stay in the intersection of those avenues rather than follow narrower paths.

Chapter 1, "Spaceflight: Discerning Its Meaning," introduces key concepts of framing, branding, and construction of meaning and then explores the heroic, pioneering spaceflight imaginary of the 1960s as an example of the power of ideas and images to shape public understanding. For Americans, human spaceflight resonates with core ideas that pervade U.S. history and culture—exploration, pioneering, the frontier, freedom, innovation, leadership, success. Establishing the origins, influences, and communication of that matrix of meaning sets up the shift into the shuttle era.

Chapter 2, "Space Shuttle: Going to Work in Space," explores the deliberate redefinition of spaceflight as practical work and routine commuting in a space truck. It identifies the verbal and visual rhetorics that NASA used to establish this concept and traces their emergence in the media as a framework for public understanding and shared meaning. A rich body of resources from the early 1970s into the mid-1980s is mined to reveal how meaning was shaped and shared to launch a new imaginary of spaceflight around a new icon for a new era.

The third chapter, "Astronauts: Reinventing the Right Stuff," examines how the astronaut as icon embodied new meanings of spaceflight. A salient distinction of the shuttle era was the broadening, diversity, and democratization of the astronaut corps through new roles and new selection criteria. The nature of the job (engineering and scientific research) contrasted with the public's ingrained perception of astronauts as pilots, especially in the wake of the two shuttle tragedies. Two memes coexisted in a shifting balance: the astronaut as exceptional and heroic, and the astronaut as an extraordinarily capable "ordinary" person.

The fourth chapter, "Science: Doing Research in Space," traces the shift in the purpose of spaceflight from practical work to laboratory research and the increase of knowledge during the 1980s and 1990s. It presents the rationale for and rapid growth of a new field—microgravity research—in life and physical sciences, and surveys selected results from shuttle science missions that helped set the stage for research on a space station. In the space station era, spaceflight became synonymous with research.

Chapter 5, "Space Station: Campaigning for a Permanent Human Presence in Space," transitions from the space shuttle as the focus of U.S. human spaceflight to NASA's push for a permanent space station from the 1980s into the new century. The space station became the new icon for justifying humans living and working off the planet. The focus here is the constant effort to shape and reshape both the rationale for the station and its actual configuration in the face of mounting opposition. Two phrases served to reshape the meaning of spaceflight once a space station claimed the agenda: "the next logical step" and "a permanent presence in space."

The sixth chapter, "Plans: Envisioning the Future in Space," surveys the episodic effort to redefine the purpose and chart the course of future human spaceflight beyond the space station. It examines the effort by presidents, NASA planners, and blue-ribbon commissions to present energizing ideas and images—to generate a new imaginary—for expanding (or curbing) the human presence in space. These exercises in charting a way into the future typically failed, in part because they were ineffectively framed for consensus or political support. The current spaceflight imaginary puts humans on the moon again, or on Mars, or visiting an asteroid at some unspecified time.

The last chapter, "Memory: Preserving Meaning," considers what the end of the shuttle era meant. With the orbiters retired to museums, the International Space Station assembled, the astronaut corps dwindled, the future-oriented Constellation program canceled, and NASA's Orion spacecraft and industry's commercial space transportation still under development in 2016, the future of U.S. human spaceflight at publication time was uncertain. Prospects for new human spaceflight rationales are unsettled, but museums that preserve the relics of the shuttle era are busy shaping public memory and the meaning of the past. Might there be some constructive dialogue between future planners and past explainers?

This exploration thus roams through four decades' worth of thinking about, and struggling with, the meaning of human spaceflight. No single concept has become the foundation for a lasting consensus about why humans should or should not be sent into space. The most enduring of several imaginaries is the frontier, which resonates for many older Americans who came of age in the mid-1900s. But this imaginary may not appeal to younger generations for whom the frontier experience is a distant and troubled one or whose entertainment choices are fantasy computer games, not pioneer tales. It may be time to step outside the box of familiar metaphors and propose a radical new paradigm—a millennial imaginary—that appeals to the values and traditions of twenty-first-century generations, the ones who will have to decide whether or not human spaceflight continues. Perhaps this book may contribute to its creation.

Chapter 1 Spaceflight: Discerning Its Meaning

An editorial cartoon depicts an eagle with a tear in its eye against a starry sky. Headlines announce the dawn of a new era as a winged spaceplane makes its first appearance. Crowds at a space launch wear T-shirts sporting a mission emblem that incorporates the symbol for woman. Astronauts in space grin and hold a sign marked "We deliver!" A white-suited astronaut floats alone like a satellite above the curve of the earth. These are some of the ideas, images, and icons that have conveyed the meaning of human spaceflight in recent times.

It has been more than fifty years since humanity entered the Space Age, a term coined to mark the advent of human activity beyond earth's atmosphere. Like other epochs named to organize and explain history—the Renaissance and the Industrial Age, for example—the Space Age signals technical, intellectual, and cultural changes that expand human life in new directions and dimensions. The most salient actions of the Space Age *in space* are ever-more-penetrating observation of the universe, placement of commercial and scientific satellites around the planet, robotic exploration of the solar system, and, to date, human spacefaring in earth orbit and to the moon.

It is now common to call the human spaceflight endeavor by the United States in the 1960s the Apollo era, and to describe it as the "heroic" or "golden" age in space.[1] Scholars and others have identified a cluster of related ideas and images, some attaining status as cultural icons, that shape our understanding and memory of that period: the space frontier and space race, the astronauts, the American flag and footprints on the moon, and the image of earthrise are richly evocative bearers of significance. The next period of U.S. human spaceflight, which can aptly be called the Space Shuttle Era, lacks a signature descriptive label or consensus about its significance. As the primary exponent of spaceflight, NASA works hard to influence public attitudes and understanding of this enterprise. But historians, journalists, political scientists, artists, and interested citizens also seek to discern what the continuing movement into space means, what its motivation or purpose is, how it affects humanity, and what its future may be.

Spaceflight is an invention. Nothing about it is natural, except perhaps the urge to explore. It is an activity first imagined and then engineered and executed at great effort and expense. Spaceflight is a cultural product of human imagination, intelligence, and will. To make sense, it needs a narrative that explains its purpose and value. To borrow a popular term in recent social and cultural studies, spaceflight is an *imaginary,* a "big idea" expressed in meaningful narratives, images, symbols, and actions that represent shared beliefs and values. Among the many imaginaries that pervade American culture with a sense of identity and shared experience are "the West," "the Melting Pot," "the American Dream," "the Cold War," and even "Democracy." A more abstract imaginary is "American Exceptionalism," the belief that the United States is unique in history and has a special destiny to spread freedom, advance technology, and ensure progress for humanity.[2]

For many Americans, the human spaceflight imaginary resonates with core ideas that pervade national history and culture: exploration, pioneering, freedom, innovation, leadership, success. The meaning of human spaceflight—the sense of its purpose and value—resides in such familiar ideas and in the images and icons that represent them. Like an ideology or a religion, spaceflight has rules and norms, traditions, rituals, a specialized language and social structure, symbols and secrets, many of them the products of belief more than necessity.[3] The linking of such ideas and human spaceflight happens so frequently in public discourse that they risk becoming clichés, widely accepted and repeated as fact. In reality, myth and metaphor are in play as well, and what seems self-evident—the "space frontier," for example—is often consciously crafted.

To ask the question "What is the meaning of human spaceflight?" is to challenge such ready answers as "the conquest of space," "pioneering the space frontier," "establishing a permanent presence in space," or "fulfilling mankind's destiny" and to probe into their origins and dissemination. Human spaceflight has itself become an imaginary rife with embedded meanings that invite interrogation and explication. Such widely understood "big ideas" are contestable.

To some extent, the meaning of human spaceflight is personal and instinctive, arising from individuals' experiences with the awesome spectacle of a launch or the shock of a space tragedy, with astronauts in public appearances, or with spacesuits and spacecraft on display in museums. But to a greater extent, the meaning of human spaceflight is deliberately framed, or invented, by its advocates and practitioners, its commentators and interpreters, and its skeptics and critics. In modern philosophy and the social sciences, it is widely posited that knowledge and reality are socially constructed, and that individuals and organizations deliberately produce and frame ideas to achieve their goals.[4] Thus, meaning can be malleable and resilient. Spaceflight is as conducive to such construction as any other reality, and as subject to message crafting and marketing as any other product, service, or institution that competes for public awareness and allegiance. Spaceflight is a product of human knowledge, beliefs, and actions.

This book delves into human spaceflight in the shuttle era to identify the ideas and images that distill its meaning and to chart their formation and transformations. This period deserves not only technical and programmatic histories (several fine volumes are in print) and popular accounts (always a staple in bookstores).[5] The shuttle era also merits penetrating attention to its ideology and iconography. Others have already characterized well the image-making and "selling" efforts to shape public perception of spaceflight in the 1960s. A comparable examination of the age of the space shuttle from the 1970s forward is warranted to reveal how key agents shaped, textured, challenged, refined, and reframed its meaning.

In the era of the space shuttle, rounded from 1970 through 2010 plus or minus a few years, the meaning of human spaceflight differed markedly from its meanings in the 1960s; it also changed in nuance throughout those four decades. These shifts in perception were not accidental; individuals within NASA, the White House and Congress, and the media thoughtfully chose words and images to re-characterize human spaceflight from its dominant prior meanings—heroic conquest of a new frontier and triumph in a space race under the banner of freedom—to new meanings for an era of routine spaceflight. The meanings of human spaceflight in the shuttle era abided in a set of ideas, images, and icons that constituted a new imaginary.

Historians bring a great variety of conceptual tools to bear on decoding such meanings. The shuttle era happened to coincide with a time of ferment in a number of social science disciplines among researchers interested in public discourse. Especially in communications, linguistics, political science, cognitive psychology, and sociology, scholars began to focus on how issues are "constructed," "framed," or "invented" for public consumption and how meaning is shared and understood. New analytical techniques flourished in research into the crafting and transmission of meaning. At the same time, communications and marketing professionals sharpened the study of brands as vessels of identity, introducing "branding" to organizations other than businesses. As these various communities of researchers discovered common interests in the influence of public rhetoric and imagery, interdisciplinary efforts arose in rhetoric, media studies, and culture studies. The study of metaphor gained new energy as a factor in public discourse, and visual culture studies of imagery gained standing. Scholars in different fields also began to investigate memory as a public and cultural phenomenon for the preservation of meaning.[6]

Spaceflight in the shuttle era is a prime candidate for examination in light of this recent scholarship. Abundant primary source evidence is available for examining the craftsmanship of meaning: in the records of those involved in the human spaceflight enterprise; in official speeches and publications; in news coverage and editorials, magazine covers, and political cartoons; and in commissioned reports and political debates. Professionals working with words tend to do so with great care, well aware that what is written or said for the public record should be carefully parsed. They discuss, debate, and negotiate intended messages, knowing that their arguments and perspectives must stand up to public scrutiny to be persuasive. Because the process of creation is typically quite deliberate, the resultant texts serve as credible evidence of intended meaning. Whether meaning is received as intended is another matter. The same is true for graphic materials that are meant to convey meaning visually in images and symbols. Decoding these sources reveals how meaning in the public sphere is created, challenged, refined, and sometimes rejected. Such analysis illuminates how recent human spaceflight was as intricately connected as the original endeavor to national myths and metaphors.

Without conducting original focus groups and opinion polls or using quantitative textual analysis techniques like social scientists do, a historian can learn from and judiciously apply concepts from such research. The analysis in this book borrows certain concepts from the disciplines mentioned and uses them to examine the meaning of human spaceflight in the shuttle era. There are, of course, fine distinctions and caveats to be made in the appropriation of such terms as discourse, framing, metaphor, myth, memory, rhetoric, and visual culture, but this lexicon has in common an emphasis on identity, values, and communication that is applicable to understanding spaceflight. Fundamentally, these fields of research demonstrate that words and images matter; they indicate and influence how we think and ultimately how we act as individuals, communities, and as a nation. The varieties of rhetoric used with purpose have the power to persuade, convince, and motivate.

In the chapters ahead, frames, metaphors, and myths are examined as aids to understanding that are created and communicated in words, images, and symbols.[7] Like a building's structural framework or a border that physically frames a painting, a conceptual framework establishes an idea and supports it. Messages are shaped for optimal appeal and usually aim to disarm or exclude contrary perspectives. Framing communicates meaning by resonance with familiar values, beliefs, and ideals; it may include metaphors, myths, and visual symbols as common forms of cultural expression. Just as the Founding Fathers are considered the framers of independence, the Constitution, and the United States government, so NASA has its framers who time and again define its goals and messages. The media relay and comment, further interpreting ideas and concepts for public understanding. The framers who shaped and reshaped the idea of human spaceflight during the shuttle era did so with resilience, even virtuosity.

Shaping the meaning of spaceflight did not begin in the space shuttle era. A powerful set of ideas, images, and icons began to emerge in the 1950s, and as elaborated in the 1960s established what spaceflight would mean during its first decade. The most prominent imaginaries from that era—conquest and frontier—point to meanings deeply rooted in resonant myths of American national identity. Human spaceflight was framed first in the context of those myths. Embodied in verbal and visual rhetoric, these ideas became the persuasive basis for public support of the grand and costly venture of sending people into space. The dominant ideas, images, and icons of early spaceflight codified meanings for that era and spread widely among Americans and the observant world.

IMAGINING THE CONQUEST OF SPACE IN THE 1950S

A framework articulated in the 1950s began to shape America's movement into space. Drawing from the vocabulary of war, imperialism, and Cold War conflict between superpowers, the "conquest of space" framed human spaceflight as travel to and from an earth-orbiting space station and eventual journeys to the moon and Mars. Conquest had dual meanings: conquering the myriad challenges of sustaining a human presence in space and also, for national security, ensuring that only a peaceful power gained the high ground of space. With World War II still fresh in memory, the Korean War in progress, rebellions erupting around the world, and the looming threat of atomic warfare, imagining spaceflight as a form of conquest reflected anxiety in tandem with hope about the future. An undercurrent of fear running through the relative peace and prosperity of America in that decade also informed dreams of space.

Technical concepts for space travel, spaceships, and space stations had appeared in print since the late nineteenth century but were not widely known until the 1950s, when a group of rocket and astronautics experts—in concert with artists, publishers, and television producer Walt Disney—introduced these ideas to the American public. First in symposia on space travel, then in a series of colorfully illustrated man-in-space articles in *Collier's* magazine and a related book titled *Across the Space Frontier*, and culminating in a series of Disney films for television, these enthusiasts presented an enticing rationale for human spaceflight and argued that it was both realistic and possible. Although the space frontier idiom also appeared then, conquest reigned as the metaphor and rhetoric for activity in space. A 1949 best seller, *The Conquest of Space*, and a 1952 *Collier's* magazine issue titled *Man Will Conquer Space Soon* set the tone (fig. 1.1).[8]

The principal architects of this framework were Wernher von Braun, who developed Germany's rockets during World War II but in the 1960s would manage NASA's development of the Saturn V launch vehicle for missions to the moon, and Willy Ley, a science writer whose popular books on rockets and space travel were well regarded. Artist Chesley Bonestell and others turned their words and ideas into stunning but believable scenes of human activity and technology in space. Together they set out to frame the reality—not a science fiction dream—of human spaceflight a full decade before the first flights into space.

The Conquest of Space, *Across the Space Frontier*, and *Collier's* presented a "blueprint of a program for the conquest of space." A space station would be a practical place for astronomical and meteorological observations, for assembling and launching other craft to the moon and beyond, and for keeping watch over the earth. Its military potential had yet greater value. A "sentinel in space" could be a reconnaissance post to help keep the peace on earth or, conversely, a fortress and battle station for waging war with guided missiles and nuclear weapons. Although Ley first popularized the "conquest of space" idiom, it was

Fig. 1.1. During the 1950s, this magazine featured the first extensive public consideration of spaceflight, reflecting a Cold War–era view of the movement into space as conquest. Smithsonian National Air and Space Museum Archives (NASM 9A05811-BW).

von Braun, the "dreamer of space, engineer of war," who touted its military significance. Embedded within the U.S. Army to manage development of guided missiles, he was astute enough to realize that a national defense rationale would outweigh scientific reasons for space travel. He assumed that the military would handle spaceflight operations and exploration like it handled the Manhattan Project that conquered the atom. Von Braun spoke of "space superiority" as the best deterrent against potential enemies, a necessity for national security in perilous times, believing that whoever first conquered space would have control over the earth.[9]

Von Braun described the shape and huge size, the orbit and assembly, and the uses of a space station. He also described in detail a three-stage rocket-plane with a returnable winged supply and crew ship that would land like an airplane, as well as small "space taxis" for use in orbit, all corollaries to the space station. Ley focused on the characteristics and operations of the space station, with attention to its provisions for all of the occupants' needs and assessments of their probable physiological and behavioral responses to weightlessness. He emphasized that "this is the blueprint for a project, not for the future but for the present. Work could start on it tomorrow" because it was all within the reach of current engineering. "What are we waiting for?" they asked.[10]

Von Braun and his fellow space travel advocates consciously set out to educate and persuade the public about the imminent reality of human spaceflight. They realized that popular support would be essential to gain resources for an ambitious space program. The well-illustrated articles, books, and television specials framed the concepts of spaceships and a space station, giving shape and realistic detail to the hardware that could extend human presence into space. Without actually using the term "spacefaring," they defined the ways and reasons for moving humanity off the planet and making space a new domain for human existence. *Collier's,* Disney, and von Braun promoted these series energetically to garner a wide audience.[11]

The images they created influenced the way Americans began to visualize spaceflight. The rocketship became iconic, appearing often in print and on screen and even becoming the centerpiece of Disney's Tomorrowland park, where enthralled visitors could take a simulated ride to the moon. It was a stocky, tapered spindle like the German V-2 war missile, but it had wings. Resembling a hybrid rocket-aircraft, it shaped early expectations for spacecraft design. Likewise, the wheel-like space station, spinning to create a semblance of gravity and fully equipped with all life-support needs, took hold and still lingers in futuristic visions. These portrayals of vehicles and spacesuits helped transform the notion of spaceflight from science fiction to science fact. Paramount Pictures came on board by releasing a motion picture based on the spaceflight imaginary invented by von Braun and his colleagues. Released in 1955, the film *The Conquest of Space* featured novel special effects meant to depict the coming reality of spaceflight, but it did not fare as well as hoped at the box office.

The blueprint for space travel and settlement that was spread out in the magazines was actually more resonant with the nation's frontier tradition of orderly movement into new territory than the abbreviated frontier concept heralded by President John F. Kennedy and the spaceflight community in the 1960s. Sometimes called the "von Braun paradigm," for the man who most eloquently and energetically promoted human spaceflight, the vision for "Crossing the Last Frontier" coherently framed a purpose, challenges, and necessary technologies for a long-term program of space exploration. Conquest would proceed incrementally to a large base in earth orbit, then to the moon, and then on to Mars or beyond.[12]

That this framework was not adopted at the outset was not a failure; it stayed alive in the 1960s, was revived in the 1980s, and may yet be realized someday. In the meantime, millions of people who were exposed to the magazine and television programs came to realize that space travel was no longer the stuff of science fiction. The proponents in the 1950s who educated much of the populace about the imminent, and in their view *inevitable,* reality of spaceflight prepared a receptive audience for America's entry into space.

INVENTING SPACEFLIGHT IN THE 1960S

Social historian David Farber labeled the turbulent 1960s "the age of great dreams."[13] The decade was shaped by movements that brought sweeping changes in American society and culture. The civil rights, equal rights, feminist, and antiwar movements; the counterculture; and the "war on poverty" challenged old notions of social order and prevailing norms, and they demanded that the cherished principles of democracy be better practiced for a more just and equitable America. Former consensus about *the* national character or *the* American experience shattered as it became evident that Americans were not, in fact, one community of one mind. The 1960s also suffered the nightmares of civil unrest; the long, unpopular war in Vietnam; and a seeming decline in morality. The presumed "climate of opinion" in American culture shifted and fractured.

In this context, human spaceflight was one of the "great dreams." The high-stakes Cold War offered a timely incentive to begin a new movement—into space. The political leadership of the United States, NASA, and the media quickly invented what had been imagined in the 1950s. They carefully framed this ambitious and expensive endeavor to be bold and unifying, and to win political and popular support, by linking it to familiar ideas and powerful images, some of which became lasting icons of the first heroic era in space. These efforts exemplify the social construction of meaning and the formation of a new imaginary.

Two cultural narratives became the basis for the "big idea" of human spaceflight: pioneering the frontier and a heroic contest with an adversary. Both held danger and extraordinary challenges, and both were knit into the intellectual and emotional fabric of the nation. The frontier narrative resonated with the themes of exploration, adventure, conquest, and the advance of civilization in America's expansion through the West, while the triumph-in-conflict narrative, recast as a "space race," situated spaceflight in the Cold War competition with the Soviet Union. Both versions related to the genesis myth of the United States as the nation founded to escape tyranny and establish liberty.[14]

The idea of space as the next great frontier arose in Americans' consciousness not long after the frontier of the West allegedly closed. The frontier myth or frontier thesis as expressed by historian Frederick Jackson Turner in 1893 was for much of the twentieth century a coherent imaginary for American culture. According to this interpretation, Americans' expansion into open space along a receding frontier and the spread of civilization into the wilderness gave the nation its distinctive character and identity. The relentless advance beyond the frontier shaped democracy and its civic values: freedom, opportunity, individualism, self-reliance, optimism, and progress. The frontier, in this idealized view, embodied exploration and adventure. Conquest of the frontier meant the rebirth of civilization at each new outpost. For more than a generation, many historians thought the frontier was the key to understanding America.[15]

The frontier was itself embedded in a grander imaginary of America as "Exploration's Nation." Exploration and discovery were such distinctive traits in the history of the United States, from earliest landfalls in the New World to twentieth-century expeditions to the polar regions, that once it appeared technically feasible to go there, space would inevitably become the next frontier. If "America is the product of an Age of Discovery that never really ended" and "the explorer therefore stands as a kind of archetypal American," as historian William Goetzmann has claimed, the pull of space as a frontier would be irresistible. Exploring new frontiers was the central act in America's continuing redefinition.[16]

It is not surprising that Americans easily accepted the frontier as a metaphor for space exploration. In the 1950s and 1960s, the West was everywhere in popular culture: scores of radio and television programs, movies, and paperback novels celebrated the frontier virtues of rugged individualism in settling new lands and the challenges of spreading law, order, and civilization. Most Westerns pitted good guys versus bad guys, blending the heroic narrative into the pioneering narrative. The entertainment environment of this era was steeped in the adventures of Wild Bill Hickok and Annie Oakley, Davy Crockett, the Lone Ranger, and Daniel Boone. *Bonanza, Lawman, Wagon Train,* and many other Western television shows portrayed versions of the frontier and the hero. Actor John Wayne became the icon for the frontier experience in movie theaters, and award-winning popular films shaped viewers' understanding of "how the west was won"—a film title suggesting both pioneering and victory. Toy stores displayed guns, stick horses, wide-brim hats, and plastic figures for playing "cowboys and Indians," and Willa Cather's tales of pioneering circulated widely in schools and libraries.

Against this popular backdrop, the principal architects of the space frontier framework in the 1960s worked in the White House, NASA, and the media. They invented the meaning of spaceflight at its genesis. Each institution has an influential role in public discourse about space policy. A president, in the role of communicator-in-chief, works with advisers to craft convincing public policy messages using rhetoric as the art of persuasion; presidential rhetoric thus indicates how issues are meant to be perceived. NASA likewise has a communications role and fosters a range of public relations activities to explain and garner support for ("sell") its programs using verbal and visual rhetoric, among other tools. The media, through reporting and editorializing, have an agenda-setting role that intersects with these and other sources of information; they frame issues and use rhetoric in response to or reaction against other agendas. During the 1960s, each of these entities actively shaped the meaning of spaceflight.[17]

When John F. Kennedy, a student of history, appropriated the frontier as the theme of his campaign and presidency, he formally established his framework for national policy. Accepting the Democratic Party's nomination in 1960, he described a "New Frontier," defining it as a set of challenges in uncharted areas of science and space, and in unsolved problems of war, peace, and social issues. He invited Americans to be pioneers on this new frontier that, like the old one, would demand their courage, innovation, and imagination. When Kennedy then urged America to enter the "new frontier of human adventure" by sending Americans into space in the 1960s, he embedded the call in a familiar framework of meaning. Whether citing the great era of seafaring exploration and calling space a vast new ocean or referring to the exploration of the West, he knew that the public understood these allusions and the broader narratives they stood for. Kennedy framed his vision in terms that would resonate with and appeal to those whose support he needed.[18]

The space frontier was not a novel concept, having occasionally appeared in the media during the 1950s, and the influential 1945 report titled *Science, The Endless Frontier* had set an earlier precedent for appropriating the frontier to argue for government-sponsored research. A 1957 editorial cartoon in the *New York Times* depicted "man's quest for knowledge" as "The New Frontier," and the next year the cartoonist John Fischetti depicted the new U.S. space program as a Conestoga wagon heading into space. The *Washington Post*'s cartoonist Herblock drew the "New Frontier Space Program" as a Conestoga rocket in 1960 before Kennedy was inaugurated. However, the term "space frontier" at first usually referred to science, especially astronomy, and rarely to human exploration beyond earth.[19]

President Kennedy elevated the concept of the space frontier to national attention by making human spaceflight a signature endeavor in his New Frontier legislative agenda. In a special Address to Congress just four months after his first State of the Union Address, he presented his list of urgent national needs, culminating in an accelerated effort to move into space. This speech was a tour de force in framing spaceflight as the new frontier. His call to achieve the goal of landing a man on the moon and returning him safely within the decade was the dramatic high note in a crescendo of initiatives, yet the single sentence itself was direct, and unembellished.[20]

The deliberations of the president and his counselors in reaching the decision to go to the moon have been recorded, but how the endeavor was presented to the public bears rhetorical analysis. How would Kennedy rally the nation to support a risky and costly challenge? His speechwriters and advisers reported that Kennedy was deeply involved in the art of writing; he discussed his main ideas with them and then edited and re-edited their drafts with an eye for the words and structure and an ear for the rhythms of the message, insisting on precision in substance and style. Chief speechwriter Ted Sorensen called Kennedy a talented writer who "believed in the power and glory of words" and "consistently took care to choose the right words in the right order that would send the right message."[21]

Together the president and his team masterfully placed the frontier frame of reference within an even broader and more persuasive frame—the defense of freedom. In stirring evocations of the nation's core identity, Kennedy stated that "we stand for freedom" and "our strength as well as our convictions have imposed upon the nation the role of leader in freedom's cause." Now spaceflight transcended pioneering and exploration on the frontier for a grander purpose: it became a moral and patriotic imperative to serve freedom. By situating spaceflight within the frames of the nation's frontier heritage and its commitment to freedom, President Kennedy tapped into two deep wells of American identity.[22]

Kennedy's "Urgent National Needs" address to Congress occurred just six weeks after Yuri Gagarin's stunning orbital flight and twenty days after Alan Shepard's fifteen-minute suborbital arc into space. The president noted the Soviet Union's head start in space but only obliquely framed spaceflight as a race. Instead, he emphasized that *free* men must fully share the movement into space, and he appealed to national pride: "For while we cannot guarantee that we shall one day be first, we can guarantee that any failure to make this effort will make us last." Kennedy gave the idea of a race new meaning as a leadership contest for the future of freedom. He more explicitly framed spaceflight as a race in a 1962 speech in Houston, Texas. "The exploration of space will go ahead. . . . We mean to be a part of it—we mean to lead it." American leadership was required because "the vows of this Nation can only be fulfilled if we in this Nation are first, and, therefore, we intend to be first."[23]

As the president framed it, the space race was not a blunt competition with America's Cold War adversary for a discrete first in space. It was a nuanced competition between the American way of life and its antithesis, a competition to win the hearts and minds of people around the world to the cause of freedom. In the political environment of the Cold War, human spaceflight could be justified as an urgent but peaceable entry in the competition for leadership, prestige, and influence. President Kennedy left no doubt that the United States must win. Linking the future to the ideals and values of the past, he rhetorically challenged the nation to enter the space age as leaders.

Like the president, the news media strongly influence what people think about current events and public policy. As the space age dawned, journalists helped frame the meaning of the daring new endeavor of spaceflight. The metaphors of the space frontier and a race for freedom's sake rapidly appeared in public parlance and spread their intended messages to broad audiences. Both rhetorical constructs became conventional wisdom. In general, media coverage accented the space race more than the frontier, perhaps because the racing metaphor enticed with a sharper sense of urgency and drama.

Newspapers had used the term "race" in the 1950s as shorthand for concern over the development of ballistic missiles and nuclear weapons, and editorial cartoonists in influential newspapers had depicted the metaphor graphically. Who was ahead in the race for missiles and arms that would travel through space? Since the October 1957 launch of the Soviets' first Sputnik satellite, however, "space race" became the metaphor for a more generalized competition to put satellites, weapons, and people in space, and even to aim for the moon. *TIME* magazine presaged the race to the moon in a whimsical 1959 cover illustration of the "man in the moon" startled by circling spacecraft and cameras under the serious banner "Space Exploration: U.S. v. Russia." "Space race" also came to mean the propaganda value of achievements in space as indicators of leadership and superiority.[24]

Editors and columnists drummed the ideological aspect of the space race. With Premier Nikita Khrushchev broadcasting that the Soviet feat in space was a clear triumph of socialism over capitalism and a victory for Communist ideology, America's opinion leaders raised the specter of the conquest of space leading to world conquest. Transforming "race" into "conquest" echoed the 1950s concerns about the national security value of spaceflight. Defeat in the space race was to be feared not simply as a loss of national prestige, but also as the possible collapse of national security and identity. Kennedy avoided the militaristic-sounding "conquest of space," but by framing spaceflight as crucial to the defense of freedom, he clearly acknowledged the threat.[25]

In reports during the week of Kennedy's May 1961 address to Congress, the national press distilled the primary messages. The *Washington Post* headline interpreted the president's message as "U.S. Is Going All-Out to Win Space Race,

Land on Moon in '67." Kennedy did not explicitly call for winning a race or beating the Soviets, but the newsmen read his meaning. *Post* articles for several days focused on various goals, among them "beating the Reds" (Communists) and boosting national prestige. The newspaper's editorial cartoonist Herblock drew a space-suited Kennedy striding toward a congressional gas pump, pointing to his accelerated space program rocket and saying, "Fill 'Er Up—I'm in a Race."[26]

The *New York Times* headlined the "Address to Congress on U.S. Role in Struggle for Freedom." Articles focused on the Cold War situation and the president's call for decisive goals and national commitments in response to the gravity of the times. Delivered just days before the president would travel abroad for his first meeting with Khrushchev, this newspaper interpreted the speech as a clear signal of U.S. determination at a time of increasing tension in relations with the Soviet Union. The *Times*' editorial response was favorable: "But it is in the spirit of free men, and the cherished traditions of our people, to accept the challenge and meet it with all our resources, material, intellectual and spiritual." This newspaper also noted general approval of the goals in Congress but wariness about the cost of the president's various initiatives.[27]

Within a few days the *Washington Post* published results of a Gallup poll taken the day before the president's address. Word had leaked about the man on the moon initiative; in fact, four days before the speech the *Post* ran a story under the headline "U.S. to Race Russians to Moon." Gallup pollsters conducted a nationwide survey on this question: "It has been estimated that it would cost the United States 40 billion dollars—or an average of about $225 per person—to send a man to the moon. Would you like to see this amount spent for this purpose, or not?" The results were 58 percent opposed, 33 percent favored, 9 percent undecided, before the public had heard the president on the matter—not an auspicious climate of opinion. Might the numbers have been different if the poll had presented the question in the president's rhetoric, situating spaceflight in the defense of freedom, a cause that resonated with America's heritage from the Revolution through World War II?[28]

A more telling challenge to the president's framing of the meaning of spaceflight appeared in *The Nation* in early June under the title "Stuntsmanship." This editorial judged the "Urgent National Needs" address "a dud" despite the "patriotic flavor" of the call for American leadership in the space race and advancing freedom. The piece charged that the president "gambled most heavily on stuntsmanship," but already opinion was turning on the probable cost of the race to the moon.[29]

That critical assessment proved unwarranted initially. By the end of the 1961 congressional session, the *New York Times* reported that Congress had given the New Frontier legislative agenda a cool reception except for the military and space initiatives. The president's own party had rebuffed the appeal to be pioneers on a broad social policy frontier, but bipartisan support had prevailed on the accelerated space program. On that issue Kennedy had persuasively framed the goal and the urgent need.[30]

In the Sputnik aftermath, opinion leaders had called for the United States to shrug off its complacency and enter the space age with a serious commitment to move ahead of the Soviets in the space race for reasons of national security and international prestige. Coupled with a growing concern about a missile gap between the United States and the Soviet Union, the space race came to be seen as a Cold War surrogate for armed competition. Democrat Kennedy had drawn a sharp distinction between himself and the Republican president Dwight Eisenhower and candidate Richard Nixon on this issue in the 1960 election: Kennedy argued that the United States should enter the race to win. Analysts had argued that the real value of the space race was psychological more than scientific or military. Although President Kennedy did not originate the space-race concept in 1961, he understood its power for committing the nation to a goal.[31]

Therefore, in the early 1960s, the rhetoric from the White House and from the mass media effectively framed America's entry into space as movement into a new frontier and as a race for freedom's sake. Space became the next arena for America to fulfill its destiny, where the nation would open new territory for exploration and also maintain its prestige as the leader of the free world—not simply by sending satellites and scientific instruments but more importantly by sending men. By framing the meaning of human spaceflight within appealing myths and traditions of American culture, the president, aided by the media, garnered enough support to launch the nation's bold and expensive venture into space.

President Kennedy cautioned that the challenge of putting a man on the moon would involve everyone; NASA could not do it alone.[32] NASA's leadership moved quickly to nurture a favorable relationship with the public. To enable everyone to understand the human spaceflight endeavor and their stake in it, the agency cultivated a close relationship with the media and produced a flood of information and messages for their consumption. This approach gave NASA an extraordinarily influential role in shaping public perception of the meaning of spaceflight.

Spaceflight held obvious journalistic appeal; one of its media masters noted that it "offered suspense, danger, colorful personalities . . . dramatic examples of human courage and skill, marvelous new machines . . . appeal to national pride, and incredible graphics." The media, especially television, offered nationwide and international exposure, world-class reporters, and masterful storytelling. NASA and the news media partnered well in framing spaceflight for the period.[33]

NASA administrator James Webb and his director of public affairs Julian Scheer "understood the powerful position NASA would command if it could convince the American people that they owned the space program and that they shared in NASA's triumphs. . . . Agency leaders wanted the American taxpayers to share in the adventure by becoming knowledgeable and supportive of the space effort." To make this connection to the public yet maintain control over its image, NASA launched an aggressive public relations operation to make it easy for the media to cover spaceflight. The agency issued news releases, press kits, fact sheets, photographs, and other helpful materials; scheduled briefings and interviews; and set up well-equipped press sites and broadcast studios to serve the needs of the media. By providing ready access to its activities, NASA cultivated the media's interest, and the media reciprocated by keeping space in the news. NASA also established its own radio and television programs, sent speakers and traveling exhibits out to communities, and produced movies and educational publications to reach the public directly.[34]

NASA's public affairs operation served the agency, the media, and the public. The agency built its image as *the* place for innovation and pioneering in science and technology and for making history by accomplishing the seemingly impossible. It cultivated the image of the astronauts as the emblem of NASA's competence, relishing their popularity and unexpected celebrity status. News organizations generally benefited from NASA's open-door public relations policy of granting equal and immediate information to all, with no favoritism. And by conducting and reporting the space program openly, NASA and the media built an interested and supportive public constituency.

Spaceflight in the 1960s produced abundant imagery thanks to NASA's penchant for documenting its activities, the media's high level of interest, and an incipient awareness of the power of photography to capture public attention. Imagery made everyone a vicarious participant in the space race and a pioneer on the space frontier. Furthermore, images quite literally framed meaning within a field separate from words. The visual content of that field stirred associations that viewers could connect to cultural myths and values without explanatory text. Several images from the 1960s era of human spaceflight achieved iconic status, communicating telegraphically what it meant to be space pioneers and to win the race to the moon. The visual rhetoric of the period complemented—and sometimes outshone—the verbal.

From their selection in 1959 as the first astronauts, the Mercury Seven epitomized the meaning of human spaceflight. They would be the ones to pioneer the way into the space frontier and carry the banner of freedom there. Originally introduced to the public as seven clean-cut young American pilots in suits and ties, they were soon photographed more deliberately as symbols of a new breed: spacemen. In 1960 a NASA photographer staged a group portrait of the seven astronauts clothed from head to toe in shiny silver spacesuits (fig. 1.2). Although functionally similar to standard olive-drab high-altitude pressure suits worn by test pilots, these training suits for the planned Mercury missions had a much different aesthetic. Posing together as in a squadron group shot, the astronauts indeed looked like spacemen, or like knights helmeted and gloved

Fig. 1.2. The first American astronauts introduced in 1959 became, in the public eye, instant heroes and icons for the space age. NASA (GPN-2000-000651) and Smithsonian National Air and Space Museum (NASM-2004-28035).

for the conquest of space. Astronauts now looked different from ordinary pilots; this portrait turned them into icons before they launched into space. Swathed in silver, almost identically posed, barely distinguishable but for their eyes, any or all of these seven embodied the meaning of human spaceflight during the first years of America's space program.[35]

For this first group of astronauts, NASA and *LIFE* magazine entered into a contract to publish a series of articles acquainting the public with the brave spacemen and their families. The astronauts were carefully scripted and photographed under the watchful eye of NASA public relations officers to appear as model Americans, suitable heroes for the space age. In the first years of the human spaceflight effort, with the benefit of publicity in one of the most popular magazines, astronauts became the icon of America's aspirations.[36]

As the space program progressed from the Mercury and Gemini missions orbiting earth to the Apollo missions to the moon, another icon took shape. Winning the race to the moon depended on fielding a new rocket more powerful than any yet in existence. For the United States, it was the awesome Saturn V, developed specifically to launch Americans to the moon. Often compared to the Statue of Liberty in height, it metaphorically launched liberty into space, in accord with President Kennedy's definition of the space race as the defense of freedom.

Photo and motion picture images of the Saturn V igniting and rising on seven and a half million pounds of thrust had a fearsome beauty. The brilliant plume and the thundering steady ascent against the grip of gravity conveyed immense power. The mighty Saturn V stood as an icon of America's power, determination, and technological prowess. Indeed, it was this technology that clinched the win in the space race, for the Soviet N-1 moon rocket never had a successful launch. Two classic launch images vividly framed the meaning of spaceflight: the Saturn V streaking past a star-spangled banner on the first lunar landing mission (Apollo 11), and the dazzling nighttime launch of the last lunar mission (Apollo 17). In both, the huge rocket seemed to rise like a new sun, shining with America's success.[37]

In December 1968, just before the Apollo 8 crew left on the historic first journey to the moon and amid rumors that the Soviets were about to attempt the same feat, the cover of *TIME* magazine captured the essence of the space race (fig. 1.3). Two figures clad in spacesuits, with the U.S. flag on one's shoulder and the Soviet red star on the other's, were bolting full-stride toward the moon. Titled "Race for the Moon," the image was just ambiguous enough: which racer was ahead? This image distilled the essence of the competition, with all elements reduced to their most basic forms: two competitors, one goal. The ideological

Fig. 1.3. This 1968 *TIME* magazine cover image captured the idea, urgency, and symbolism of the space race at a point when the race was too close to call. From the pages of TIME. TIME and the TIME logo are registered trademarks of Time Inc. used under license. © 1968 Time Inc. All rights reserved. Licensed from TIME and published with permission of Time Inc. Reproduction in any manner in any language in whole or in part without written permission is prohibited.

competition was likewise reduced to the two national symbols—flag and star—not the most noticeable elements in the composition. In a geopolitical context, the image read as suspense before imminent victory, but without that context, the image almost suggested that in a photo finish, who won the race might be less important than that humankind had accomplished something amazing.[38]

Of course, it did matter which nation reached the moon first, and the first images became permanent records of that triumph. Two images from the Apollo 11 mission became especially iconic for the meaning of human spaceflight as pioneering on the frontier and winning the race. One was the image of an astronaut alone on the barren lunar landscape—Buzz Aldrin, left arm bent so the U.S. flag on his shoulder was clearly visible. To the discerning eye, Neil Armstrong was reflected in Aldrin's helmet visor as he took the picture, but that detail was almost lost in the composition that featured a sole being on a strange world with only his shadow and footprints for company. The sense of isolation and otherworldliness was palpable, and the image clearly stated, "I am here on the moon."

Another Apollo 11 image more blatantly made the point "America is here on the moon." Buzz Aldrin posed near the United States flag "flying" from a pole planted on the moon. Man, moon, U.S. flag—these three essential icons conveyed the message. Each of the six Apollo landing crews enacted this ancient ritual of explorers, not literally to claim the moon as territory but symbolically to prove the priority of America's arrival there. These two iconic images marked both the realization of entering the space frontier and the triumphant end of the space race.[39]

TIME celebrated the Apollo 11 landing on the moon in its regular weekly issue printed before the actual mission photos were available. The cover art for "Man on the Moon" featured a scene very like the actual flag-on-the-moon photograph that soon was published. On the cover, an astronaut standing on the moon held a flag on a pole as if preparing to raise it in place. The astronaut's visor was up to reveal the face of Neil Armstrong, who ironically was not photographed in that manner while on the moon. The large flag dominated the scene. As in the actual photograph, the essential elements—man, moon, and flag—spoke to both the historic event and to the myths that gave it meaning.[40]

LIFE magazine used two other images on its covers just after the Apollo 11 mission. The regular weekly issue featured a view looking down from the lander on the flag and the footprint-trampled lunar surface. With no astronaut in view, the image only obliquely acknowledged the feat (and feet) of men on the moon. The cover of a special issue, *To the Moon and Back,* featured a close-cropped version of the Aldrin solo image enlarged to dominate the cover, make Armstrong visible in the visor reflection, and accent the flag on the shoulder of the spacesuit. Like the *TIME* covers, these framed the meaning of spaceflight with the three essential icons: man, moon, and flag. Both the *TIME* and *LIFE* magazine man-on-the-moon issues became instant collector's items for capturing in images this epic narrative of human achievement.[41]

Perhaps the most surprising images to become icons of the Apollo era of human spaceflight were two photographs of earth: the Apollo 8 image called "Earthrise" (1968) and the Apollo 17 image called "Whole Earth" (1972). Astronauts took these photos on the first and last missions to the moon. "Earthrise," taken in lunar orbit, seemed to convey the idea of the space frontier more than many other images taken on the surface, because it reversed humanity's perspective on where home was and what spaceflight really meant as a voyage *away from,* not only a voyage *to,* somewhere. The partial earth visible beyond the curved horizon of the moon now clearly was a distant planet; the astronauts had indeed left home and crossed the space frontier to a new place. This image came to bear poetic significance, revealing earth as an oasis in space, an island of life, a blue jewel in the dark void. The space frontier was no longer an abstraction. It, like earth, had become part of the sphere of human life.[42]

The "Whole Earth" image from Apollo 17 (actually a half-earth image, but a full disk) came to carry a different set of meanings that had less to do with where humanity was in space but how humanity should steward its own spaceship earth. Adopted as an icon by various environmental and other causes, this image came to bear meanings tangential to human spaceflight. Few images obtained from spaceflight have had as much currency as this as a logo for activist, globalist, and commercial uses on earth.[43]

Images that are widely published and widely recognized become iconic when they are so readily understood they need no explanation. Even so, they can be "unpacked" or explicated as a cultural text to reveal more of their backstory and context. Viewers bring their own experiences to interpreting images; their own knowledge, beliefs, attitudes, and identity—their gaze—affect the emotional power of images beyond their content and aesthetic. The visual culture of the early space age was so novel and fascinating that it made profound and lasting impressions. Although nothing like some of these images had been seen before, some of them resonated with the frontier and conquest myths and thus had meaning beyond the moment. They became the iconography of the space age. More than fifty years later, some of these just mentioned still rank among the most popular iconic images of the twentieth century.[44]

CRITIQUES OF THE SPACE RACE AND SPACE FRONTIER

Of course, the space frontier and space race did not engender uniform consensus. Skeptics and critics of the venture into human spaceflight responded with contrary views and framed persuasive arguments to make their case. It was evident immediately after President Kennedy's special address to Congress in 1961 that some with power to affect the course of events had reservations, if not outright opposition.

The most noticeable early critical voice was former general and president Dwight Eisenhower, on whose watch the Soviets launched Sputnik and Lunik satellites. He had declined to be pressured into a race that he considered a gimmick, arguing instead that the nation should proceed into space according to an orderly plan to meet valid national security and scientific objectives. He thought it unnecessary to rush and inappropriate for the U.S. agenda to be set by external events. As the senior Republican, he spoke harshly about the new Democratic administration's crash program to achieve a space spectacle and the high cost of the effort, even going so far as to brand it "nuts."[45]

Republican leaders in Congress, joined by some Democrats, objected to the space race on grounds of its cost. Wary of multi-billion-dollar expenditures without tax increases to pay for the accelerated space program, these critics sounded an early alarm in 1961 that gained enough support by 1963 to start curbing the NASA budget. They did not reframe the goal of the space program itself but instead placed the space race within the economic sphere of inflation and inadequately funded social welfare programs, urging that national priorities be reexamined.[46]

Newspapers reported the political-economic critique of the space race, but it did not gain much traction on influential editorial pages. Opinion pieces in the *New York Times* and the *Washington Post* mentioned critics' misgivings about the cost of the effort to send men to the moon and the burden on American taxpayers but maintained that winning the race was essential and worth the price. An early editorial piece anticipated such a critique: "There is an acute awareness of the cost of space exploration . . . taxpayers and politicians are likely to denounce the proposed flight to the moon as a gigantic boondoggle . . . [and make] complaints . . . about frittering away billions in space." The writer then dismissed these concerns by stating that spending "vast sums on a race to the Moon . . . is one of the stark necessities of the age." In general, opinion leaders did not encourage the "haste makes waste" or "spending money to better advantage on earth" arguments.[47]

Many of the critical editorial cartoons during the mid- and late 1960s shared a common theme, juxtaposing the cost of the space program with unmet social needs on earth. Such drawings notably appeared in 1969 at the same time that other cartoonists celebrated the first lunar landing and the space-race triumph. These critiques framed the space program as a misplaced priority or a funding hog that deprived other worthy causes of attention. In the cartoons, bags of money for space or skyrocketing costs were offset by the poor, urban decay, education and health care needs, pollution, disarmament, war, and other ills. One news cartoonist marked the last Apollo mission in 1972 with a succinct message: "It's been great—but we can use the money for other things!" while another used the tagline "one small step for a man, one giant leap for the taxpayer."[48]

The scientific community also was skeptical of the space race; the reluctance might seem somewhat a paradox as human spaceflight poured funding into science and technology sectors. Yet a 1963 poll of twenty-five of fifty-five Nobel Prize winners in the United States objected to the crash effort to win a race to the moon and advised a more deliberate pace. Their reasons were varied. Admittedly, scientific exploration was not the primary objective of the Apollo program, so many scientists preferred a more research-based approach. Others were troubled by the arbitrary deadline or the priority claim on the nation's resources. Members of professions governed by scientific facts and logic understandably were wary of rhetoric.[49]

The momentum of such arguments grew strong enough that critical opinion pieces appeared in *Science* and other journals. These comments noted that skepticism was growing, that "glittering but tinny arguments" did not satisfy as the rationale for a big space program, and that it was doubtful the purported terrestrial benefits and products from spaceflight would materialize. The conclusion was that "an accelerated moon program can command political support only in a Cold War context," without which it would be much more difficult to sell to the public and Congress. Although a joint NASA–National Academy of Sciences study on space sciences endorsed NASA's scientific program, a lingering skeptical attitude ensured that any commitments to a post-Apollo space program would be debated.[50]

The Moon-Doggle, a book published in 1964, presented a biting scientific critique of the race to the moon as "a monumental misdecision." The sociologist author protested the mismanagement of science and the impulsive decision to go to the moon in what he labeled a "cash and crash" program. He argued that too many areas of science were malnourished while human spaceflight consumed billions of dollars, and the headlong pace of a crash program was unprofessional. Treating the moon as a status symbol to be won in a race begged a critical question: How important, really, was prestige? He offered that instead of spurious arguments to support a hastily considered space race, a national science policy with carefully set goals and guidelines should be in place for future decisions of this scope.[51]

In general, the thrust of the scientific critique did not challenge the space frontier concept but did challenge all dimensions of the space race. Some vocal representatives of the scientific community were not persuaded by the economic or scientific aspects of the space race. They did not attempt to reframe human spaceflight into a more compelling concept. Instead, they focused on the decision-making process and argued that there was a better way to make such national commitments.

One measure of the effectiveness of conceptual frames such as the space frontier and space race is public opinion. If the framing strategy works, it will influence public opinion favorably. It has long been a premise that public support for spaceflight was strong during the space-race years in the 1960s, although, as noted previously, a Gallup poll that sampled public opinion shortly before President Kennedy's call to land a man on the moon revealed that only one in three people favored the effort. Analyses of polling data throughout the decade have shown that public support was never as strong as 50 percent, although it was fairly steady, fluctuating between about 33 and 45 percent. When asked whether they favored government spending to send a man to the moon, the majority consistently said no. When asked whether the government should do more or less in space, a higher number always said "less," and the "less" percentage rose noticeably after the first Apollo landings. The cost of spaceflight and the pull of other societal needs were important factors in public opinion.[52]

What does this suggest about the effectiveness of framing human spaceflight as a space race or exploration of the space frontier? Ostensibly these concepts were chosen to be especially resonant with cultural beliefs and traditions, yet they did not persuade even half the public to embrace the program. When Americans stood on the moon with the U.S. flag, public interest waxed with pride in the moment, but for the rest of the 1960s it appears that the public was less enthusiastic than assumed. In polls, the space program was often cited as a candidate for funding cuts.

Analysis of the meaning of human spaceflight in that period has ranged across the disciplines of political, social, cultural, and Western history. In general, historians have placed the space race into the larger, more complex frame of the Cold War. Only in the context of a geopolitical superpower rivalry could the vast mobilization and expense of the project to send men to the moon be rationalized. The principal histories of this era attend to the political imperatives that spawned and sustained the space race.[53]

Upon its successful culmination in the Apollo 11 mission, the space race achieved the goals that President Kennedy had deemed so urgent in 1961: to demonstrate American preeminence in technology and ideology, and to regain international prestige by accomplishing a great venture under the flag of freedom in full view of the world. Conventional wisdom ranks the landing on the moon as one of the signature events in the twentieth century and indeed in human history.

Yet some historians also fault the space race as a conceptual frame for human spaceflight and argue that it actually was a disservice to the cause of space exploration because it lacked a strategy for continued endeavors in space. In this view, the space race has been dismissed as a stunt, a series of space spectaculars, and a dead end that put the future of human spaceflight in jeopardy.[54]

Scholars have also rethought the frontier myth, and spaceflight framed as pioneering has lost some traction. Prominent historians of the American West have reached different conclusions about the aptness of the frontier myth as a frame of reference for the movement into space. One has exposed the flaws of using an idealized frontier concept in the service of space exploration; doing so overlooks grave problems in the real frontier experience. Another has cast the space frontier in a third great age of discovery as a sequel to the historic seafaring and transcontinental explorations but has pointed to what spaceflight lacks, by comparison. A controversial reinterpretation of the frontier from the perspective of art history and visual culture dissected the creation and depiction of the mythic West in contrast to the brutal history of its conquest.[55] In short, the West, now understood as a more conflicted than heroic narrative, is no longer uncontested as the quintessential American myth. With hindsight, it is evident that the space race did not immediately open the space frontier to a long-term human presence. Instead, the Apollo program ended and the U.S. space community had to articulate a new purpose for further human spaceflight.

Historians of the first era of human spaceflight generally agree that it was an anomaly—a crash program launched by presidential leadership and "sold" as an urgent national imperative was not the normal way to set space policy. It was effective only because of the pressure of the Cold War environment. In that milieu, the space-frontier and space-race concepts proved persuasive. Absent such pressure, a much different approach to human spaceflight and a relaxed timetable would have been likely. In hindsight, history reveals the interdependence of frames and their contexts; timing was the critical factor in the persuasive power of the space race as a conceptual frame.

The spaceflight community adopted the frontier myth as the scholarly community was challenging it in the 1950s and 1960s. While many historians began to abandon the significance of the frontier as a flawed myth, NASA and spaceflight advocates in the media and elsewhere still found it an apt source of inspiration. Western scholar Richard Slotkin has observed that through familiarity, a myth distills into "a set of powerfully evocative and resonant 'icons' or signals that are coded with meaning. . . . The terminology of the Myth of the Frontier has become part of our common language," resistant to revision in the popular mind. Just as historians are unlikely to follow engineering scholarship, so the aerospace community launching spaceflight was probably unaware of historical criticism. As historians of the West have noted, "What has become a cliché for the scholars remains an icon for most other Americans."[56]

The frontier myth seemed to fit comfortably the early years of space exploration as humans ventured first into near-earth orbit and then pressed on to a new world, the moon. Embarking into an unknown and hostile environment, conquering this "wilderness," proving human and national courage by accepting formidable risks, welcoming challenges and defying failure, vanquishing the ideological adversary—these characteristics and more became the hallmarks of the culture of human spaceflight in the 1960s. Risk was part of the allure, and astronauts became heroes not only for their own personal courage but also as surrogate explorers for the population on the ground. They were modern pioneers in a new frontier saga. With no environment to despoil and no wildlife or natives to displace, enthusiasts could anchor America's course in space in past experience in the frontier West.

The vigor of the frontier myth for the early spaceflight community may be attributed in part to the popularity of Western movies, television programs, and fiction in the 1950s. While many of the engineers and managers who brought the space program into being were influenced by popular science fiction, the popular culture of their youth was also populated with cowboys and pioneers on the Western frontier. Quite likely they assimilated many of the images and values from America's frontier heritage from their textbooks and their entertainment. History texts like *Frontier America* or referring to "the frontier's impact on American life" were still being published at the dawn of the space age.[57]

It has been argued that the frontier was an "entitling myth" for the space age; it was appropriated to justify a course of action by appealing to a concept already familiar to the populace. Such myths can be efficient in times of social or political change, when public support is crucial for change to occur, and Kennedy's presidency was such a time of national stress and change. Aligning a new agenda with a narrative and beliefs accepted in the culture would have been a wise strategy for persuading politicians and the public to get on board.[58]

However, the space-race metaphor for human spaceflight held its own demise, for as soon as the race ended in victory there was no further point in continuing to run full speed ahead. Indeed, some judged the race over even before the landing on the moon. In 1963 Khrushchev hinted that the Soviet Union no longer set its sights on the moon, but Americans suspected this was a ruse. By 1965 U.S. achievements in space were outpacing Soviet firsts, and a scorecard published at the end of 1968 showed the United States convincingly ahead in all the metrics of human spaceflight. Moreover, by mid-decade, editorial opinion began to turn against the race, on the premise that there were more urgent needs to be met on earth than this enormously expensive contest for prestige.[59]

The United States reached the moon first, but it was hard to point to tangible consequences in the Cold War struggle for the hearts and minds of men. By the end of the Apollo era, pundits had questioned whether it really was a race and whether the Soviets long ago bowed out to save face. Sterner critics came to view the space race as a dead end that left no path to the meaningful continuation of human spaceflight. With no long-range goals and no versatile technologies adaptable to purposes other than lunar missions, there was neither mission nor means for human spaceflight.

Likewise, the frontier framework proved transitory without a program of methodical exploration and settlement. Having landed on the moon six times and done geological sampling of soil and rocks in different locales, NASA reached the end of the nation's lunar agenda. With no mandate to continue pioneering on that world, the frontier framework could not be sustained without another place to venture.

With the meaning frameworks of the 1960s collapsing, NASA and other promoters of spaceflight faced a new challenge by the end of the decade. If human spaceflight were to continue, it would be necessary to give it fresh meaning, to frame and brand it in credible new terms, to intentionally shape that meaning in order to revitalize public interest and political support. To do less would court the demise of human spaceflight.

REINVENTING SPACEFLIGHT FOR THE 1970S AND BEYOND

When Wernher von Braun, Willy Ley, Chesley Bonestell, and others in the early 1950s had envisioned a winged shuttlecraft or ferry-rocket operating in near-earth space, the enticing artwork in *Collier's* magazines and Disney films for television was not entirely imaginary. Aerospace research programs were already well under way to test the feasibility of winged and wingless flying vehicles that could return from space to an airplane-like landing. Much of that research was predicated on theoretical and experimental foundations laid since the early 1900s by rocket pioneers Konstantin Tsiolkovsky, Hermann Oberth, Robert Goddard, Max Valier, Fritz von Opel, Walter Hohmann, and most notably Eugen Sänger, who developed the hypersonic reusable winged spacecraft concept, *Silverbird,* elaborated with Irene Bredt during the 1930s and 1940s. These concepts anticipated the streamlined rocket planes that would punch through the sky in mid-century test flights.[60]

Popularizers helped prepare the public for this kind of spaceflight, but it was engineers and pilots who wrestled with myriad technical problems to make reusable winged spacecraft a reality. Through forward thinking and flight research, they laid the foundations for a new era in spaceflight well before the first era culminated in the moon landings. Long before there was a specific requirement for a reusable piloted spaceplane, the aerospace research community became intensely interested in flying faster and higher and also gliding back from high altitude to a runway landing. A series of experimental X-planes and lifting bodies worked out many of those aerodynamic issues. Some precursors of the space shuttle were already in the air, but not yet in space, in the 1940s and 1950s. By the early 1960s, some had breached the boundary of space to open the way for a winged spacecraft not yet on the drawing board.[61]

The preceding analysis of the meaning of human spaceflight in the seminal 1950s and 1960s suggests approaches for examining the meanings of spaceflight in the next era—meanings invented around the space shuttle spaceplane. The process of framing new meaning for this era began even before the first landing on the moon. We can watch it take shape in NASA documents, presidential

Fig. 1.4. *TIME* welcomed the space shuttle home from its first mission in 1981 with a red, white, and blue cover that echoed NASA's "new era" theme. At the time, "Right On!" meant *awesome* or *congratulations*. From the pages of TIME. TIME and the TIME logo are registered trademarks of Time Inc. used under license. © 1981 Time Inc. All rights reserved. Licensed from TIME and published with permission of Time Inc. Reproduction in any manner in any language in whole or in part without written permission is prohibited.

rhetoric, study group reports, media commentary, criticism, and the images and icons that emerged as human spaceflight began anew.

The following chapters examine shuttle-era human spaceflight in the promotional frameworks that have been in play during four decades of striving to articulate the justification for continued human spaceflight. Proponents had to devise compelling arguments for keeping the engineering and management expertise and the technology readiness that had become a national resource. That meant shaping new frames of meaning for the future space program—among them, routine space transportation, a permanent presence in space, a diverse astronaut corps, and international partnerships. At the same time, opponents challenged those meanings in an effort to rein in an expensive space enterprise or a program that, in their view, glorified human spaceflight at the expense of space science and robotic exploration. Since the 1960s, the human spaceflight imaginary has been a matter of considerable debate and evolution.

One image in particular epitomized the reinvention of human spaceflight: the *TIME* magazine cover published after the first space shuttle mission in 1981 (fig. 1.4). It framed the winged orbiter *Columbia* gliding home on final approach with the exclamation "Right On! Winging into a New Era." The glide slope angle of the orbiter had a dynamic effect, and there was no doubt about the meaning of the red-white-blue color scheme and the U.S. flag near the center of the image. America was doing human spaceflight again after more than five years on the ground, and this image was a visual "thumbs-up" expression of pride in the innovative and successful beginning of a new era.

Chapter 2 Space Shuttle: Going to Work in Space

After cheers and handshakes, the four astronauts aboard space shuttle *Columbia* gathered for a celebratory crew portrait. They had just completed the main task of the shuttle's first operational mission. After four orbital flight tests, NASA in mid-1982 declared the new spaceship ready for business and loaded two large commercial communications satellites into its cargo bay. The crew's job: take these satellites into low-earth orbit, carefully release them overboard, and get out of the way before their rocket stages ignited for the boost into higher orbit. This mission demonstrated why the reusable shuttle existed: to make space accessible for human activity, including work previously done by expendable launch vehicles. Upon landing on the moon, the Apollo 11 crew had unveiled an engraved plaque bearing the noble statement "We came in peace for all mankind." The beaming crew of the fifth shuttle mission also marked their historic moment, holding before the camera a hand-lettered sign with a more prosaic statement: "We Deliver."[1]

Those two words became the motif for scores of similar missions in the next thirty years. Perhaps more than any other claims made about the space shuttle, "We Deliver" signaled the new ideology of human spaceflight in the shuttle era. Going to space meant going to work. Astronauts had jobs to do for NASA and for the shuttle's customers. They delivered satellites, telescopes, and experiments into space and brought research data home. They tested new technologies devised to use space for practical purposes. They delivered evidence of the value of human beings in space for problem solving and doing complex tasks inside and outside their vehicle. Shuttle astronauts eventually completed a ten-year construction project by delivering and assembling the International Space Station, one huge piece at a time. They delivered confidence that humans belong in space. The succinct "We Deliver" motto soon appeared in other crew photos and as the title on some of NASA's shuttle marketing materials (plate 1).

With the shuttle, NASA shifted emphasis from exploration to utilization of space. The orbiter was a hybrid vehicle—part camper, part moving truck, service station, research lab, observatory, and construction rig—with wings and wheels like an airplane and engines like a rocket. Nothing like it had ever flown in space before. The shape of the shuttle orbiter—a broad-winged craft with a large cargo bay body—signaled usefulness. It was designed for doing work in near-earth orbit, landing, being serviced for its next job, and heading back into orbit on another utilitarian mission as a reusable, multipurpose spaceplane.

From the mid-1970s on, the signature theme for human spaceflight was "going to work." NASA and the media heralded this period as a new era and framed spaceflight as a normal, routine, practical, and beneficial enterprise. In sharp contrast to the high-stakes patriotic quest of the 1960s, the rationale for spaceflight in the new era accented working-class values with a more utilitarian appeal. This new imaginary spread through depictions of both the shuttle and its crews in imagery and icons sometimes associated with blue-collar labor, as astronaut repairmen made service calls in a vehicle often called a space truck.

Why did NASA deem it important to make spaceflight routine and economical? How did reusability relate to routine flight and thus to economical access to space? For what purposes would humans go there? Why did the United States want to have this capability? To answer such questions, NASA conceived a distinct ideology for human spaceflight in the space shuttle era, framed it in a new rhetoric, and created icons for its meaning. Popular media in the 1970s and 1980s aided in the public dissemination of this rationale but also targeted its weaknesses. As the shuttle became more familiar, NASA worked to keep the rationale resilient through a vocabulary and visual culture that evolved with actual experience. Novel as it was, spaceflight on the shuttle was not the ultimate imaginary, and NASA needed to sustain public interest to support a more expansive future.[2]

* * *

Even before men first stepped on the moon, debate and planning began for the space program's next steps. In early 1969 President Richard Nixon appointed a high-level Space Task Group—his vice president, the NASA administrator, the secretary of the air force, and the president's science adviser—to chart the nation's future in space. They had already begun to plot their own visions of a future in space and were prepared to influence the nation's plan. The group's September 1969 report, *The Post-Apollo Space Program: Directions for the Future,* called for a balanced approach to "manned" and "unmanned" exploration that would preserve aerospace expertise and advance technologies. Beyond that generality, the Space Task Group deliberately aimed to define the guiding purpose and framework for continued human spaceflight.

With the cancellation of the last planned Apollo missions, it was clear that a new rationale for spaceflight was necessary. In some quarters the lunar landing missions were seen as repetitive and not worth continued expenditures. Social and political critics were challenging the expense of human spaceflight as unwarranted in the face of such social inequities as poverty and racism. The war in Vietnam was sapping America's confidence as a technological superpower. The social consensus about national priorities had changed, and spaceflight slipped lower on the national agenda. For NASA to thrive in a post-Apollo world, it would need to reinvigorate its mission and identity.

The Space Task Group report held two key concepts in tension: exploration and "exploitation" of space. It called for an increased emphasis on "utilization" of space for direct practical benefits and applications that would improve the quality of life on earth, yet the group also urged the adoption of a grand, adventurous goal to animate the space program and public support: a mission to Mars by the end of the century. Toward that goal they outlined the necessary infrastructure—a space transportation system and space station operating in earth orbit—that would enable ready access to space and a springboard for exploration. To be feasible, this framework would have to be built on pillars of reusability and economy.[3]

Reusability already was factoring into high-speed, high-altitude aerospace research aircraft, but it was an untried concept for orbital flight. Common sense could make the case for lowering the cost of spaceflight and space operations. The looming challenges were to develop new capabilities for human spaceflight and drive the cost down. The Space Task Group argued that the long-range goal of a manned mission to Mars would provide the impetus for meeting those challenges.

Assuming that public and political skepticism would diminish if costs were reduced and if compelling new missions were planned, the authors wrote a prescription for framing the human spaceflight vision to win support. It should have clear, easily understood values; a long-term goal as a beacon; attainable steps for progress toward the goal; practical benefits; and both technological and inspirational challenges. Practical utilization of space was necessary but not sufficient; spaceflight must also appeal to the spirit of adventure and exploration. This advice framed a strategy for "selling" human spaceflight in the public and political arenas.

The vision of a space station and a human mission to Mars proved too grandiose and expensive for the time, but the space transportation system gained a foothold, and the guidelines in the report shaped the thinking that led to the space shuttle. Its advocates believed that a reusable space transportation system would lower the cost of spaceflight and be capable of serving the full spectrum of civilian, national security, and commercial missions. They believed, or at least proclaimed, that the new spacecraft would carry people and payloads to orbit and back as routinely as an airline.

NASA, the air force, and aerospace corporations had begun future studies before the Space Task Group formed, including studies of a transport vehicle to service a future space station. NASA launched 1969 by triggering a round of studies on a reusable spacecraft already being called, informally, a shuttle. The starting assumption was that it would be a fully reusable vehicle, but various designs and configurations were possible. The studies would identify the problems associated with each and narrow the range of possibilities. A satisfactory solution had to take into account technical, cost, schedule, operations, and maintenance factors that were not necessarily ideal for any one design, so evaluation meant making tradeoffs and compromises. The decisive factor proved to be cost, forcing a solution that cost as little as possible, either to develop or to operate.[4]

During this series of studies from 1969 into 1972, selected aerospace companies each investigated a particular concept in multiple forms. Some focused on wingless lifting bodies; others on narrow straight-wing or wide, triangular delta-wing designs, or pop-out deployable wings. The strong bias for an integrated crew and cargo vehicle precluded capsule designs. The lifting bodies and deployable wing designs, though impressive on some counts, were ruled out as unsuitable. That left only winged vehicles under consideration, and ultimately NASA chose to develop a long-body, delta-wing vehicle design.[5]

Many within the aerospace community greeted the advent of a winged spacecraft with enthusiasm. They considered a capsule's ballistic reentry and splashdown in the ocean inelegant, if not undignified; a spacecraft "should" land and be at least partially controlled by a human pilot; that is, it "should" have wings.

Flight research programs had already established the precedent of controlled gliding reentry from space, and the experimental work boosted confidence that reentry in a reusable winged spaceplane was feasible. Furthermore, if the new spacecraft were to go into routine airplane-like service and achieve airline-like economies, much about its design, maintenance, and operation could be patterned on aircraft.[6]

As the design studies progressed, NASA burnished the space shuttle's potential to ensure that it would gain favor. A classic strategy for winning support is to frame a concept for maximum buy-in; that is, making sure that something about it appeals to everyone who matters. In the case of the shuttle, the vehicle became "all things to all people." Its promised capabilities and flight rate may have stretched credibility, but they were key to the strategy of persuading the widest possible base of support. Promoting development of a space transportation system that could serve as a shuttle to a space station, deliver satellites and repair them in orbit, carry commercial payloads, serve Department of Defense needs, and support a research laboratory cultivated a broader constituency than a less capable system would. The genius of this approach also proved to be its weakness; once the shuttle really existed, its multiple purposes and expectations became a source of many of its problems.

These multipurpose roles of the space shuttle were germane to NASA's framing the rationale for human spaceflight in terms that would connect with its constituents in the political and public arenas. The reusable winged spacecraft inspired the slogan "A New Era" in spaceflight. NASA captured its essential purpose in another motto suggestive of routine activity: "Going to Work in Space." That motto became the theme for a series of posters issued after the first shuttle missions; each featured a spectacular image of the orbiter launching, in space, or returning to give the idea a strong visual expression (plate 2).

President Nixon's announcement, in January 1972, of the decision to develop a space shuttle conveyed all the tenets and keywords of the rationale for a new era in spaceflight. NASA had submitted to the White House a draft statement of four reasons the shuttle was "the right next step in manned spaceflight," but the text reflected input from other sources as well. Little of NASA's wording survived, but threads of its ideas were there. Speechwriters distilled the essence of the shuttle rationale as "an entirely new type of space transportation system designed to help transform the space frontier . . . into familiar territory, easily accessible for human endeavor. . . . It will revolutionize transportation into near space, by routinizing it. It will take the astronomical costs out of astronautics . . . delivering the rich benefits of practical space utilization . . . to achieve a real working presence in space . . . so that men and women with work to do in space can 'commute' aloft." Nixon called the shuttle the "workhorse" of the U.S. presence in space and stated that developing it was the right thing to do for its utility and the nation's space leadership role (and it would boost the economically stressed aerospace industry). With that endorsement, NASA settled into turning the ideas of this new era into reality.[7]

Apart from the daunting engineering and technology challenges of building the new spacecraft, NASA also wrestled with a more mundane yet still important problem: what to call it. Although the vehicle had been informally called a shuttle for years, some within the agency felt that the name was too ordinary. The deputy administrator, NASA's point man to Congress and the White House, expressed concern that with this name, "it is clear that we are not communicating what it [the shuttle] is and what it does." Others shared the sentiment that the name of the reusable spacecraft should better capture the spirit of the new era and the intended revolution in spaceflight.[8]

For several years, NASA empowered committees to solicit and review suggested names and forward recommendations to the administrator. People throughout the agency weighed in on their own or in response to a public affairs office questionnaire. Among the names considered appropriate were Spaceclipper or Skyclipper, Astroship, Astroplane, Spaceliner, Spaceplane, and Spacemobile. Suggestions for a mythological name were dismissed because "such names are insufficiently utilitarian and may lead the public to associate the shuttle in an undesirable way with the more costly Apollo program." A public affairs officer hoped for "an exciting and appropriate name. 'Shuttle' is really very descriptive, but certainly does not have a space age ring." Another public affairs officer made the case for shuttle as "a very descriptive name . . . exactly what we want—a vehicle that goes to and from space . . . regularly, on quick turnaround and economically." He added wryly that "work horses usually don't get dramatic names."[9]

The effort ended in 1976 without a consensus on a more satisfactory name. By then the term "shuttle" had been in use so long that the moment had passed to refine its identity without causing confusion. Also by then, Rockwell International had assembled the first shuttle vehicle and was ready to reveal it.

A crowd of government and aerospace dignitaries and a contingent of news reporters gathered in anticipation on a sunny morning in the Antelope Valley desert north of Los Angeles. It was Friday, September 17, 1976, and they had come on a pilgrimage to Palmdale from Washington, D.C., Florida, Texas, and points in California and elsewhere. The reason for the journey: to witness the debut of America's next spacecraft.

A signal of something new was the presence of the producer and cast of the *Star Trek* television show and the military band's inclusion of the *Star Trek* theme in the musical program. Another novelty was the name *Enterprise* freshly painted on the side of the huge spacecraft, as large as a commercial jet airplane. Neither of these features was part of NASA's original public relations plan. They had cropped up just days earlier when President Gerald Ford, influenced by petitions from *Star Trek* fans asking that the vehicle be named for the show's starship *Enterprise,* recommended the name to NASA. The space agency's own preference, after much internal discussion, had been *Constitution,* in honor of the U.S. bicentennial year (1976) and Constitution Day (September 17), to brand the new spacecraft with a name steeped in American heritage. Now, instead, it was adding a Hollywood-style aura to the occasion and introducing a vehicle name plucked from popular entertainment.[10]

This day not only marked the rollout of the world's first reusable spacecraft. NASA also focused attention on the new rationale for human spaceflight, framing the day as the beginning of a new era. The basic themes of the space shuttle announcement in 1972, now better developed and polished, gave meaning to the festivities in an admirable exercise in message management. The matter-of-fact NASA press kit for the shuttle rollout event contained no hyperbole, but the opening line had a key phrase: "the new era of space transportation." That phrase appeared in almost every news report from the event, effectively putting the space shuttle in a frame big enough to accommodate all manner of "new."[11]

Despite the gloss of Hollywood celebrity, the themes of the rollout event were practical. NASA characterized the space shuttle as "a versatile and reusable spacecraft," and the orbiter as the "workhorse" of the new space transportation system. Its varied missions would include placing satellites into orbit, retrieving and repairing or returning malfunctioning satellites, carrying the reusable Spacelab scientific laboratory for researchers to use in space, and deploying probes with upper stages to propel them to planetary destinations. To make the point of reusability, the press kit noted that the orbiter was designed to be used as many as a hundred times—a simply stated but astonishing claim.[12]

All the speakers for the event stayed on message, following NASA's cues. Administrator James Fletcher and head of spaceflight John Yardley spoke of the new era of entering space permanently and usefully. Speakers from the congressional space committees, Representative Olin "Tiger" Teague and Senator Barry Goldwater, hailed the new era of routine access to space in an economical space transportation system. Willard Rockwell, chief executive officer of Rockwell International, the space shuttle prime contractor, also spoke of the new chapter in history and the productive use of space.[13]

The unveiling focused attention on the vehicle's capability with little mention of human spaceflight per se. Surely some astronauts attended, but apparently none spoke or gave interviews. The remarks and print materials emphasized the reusable spacecraft more than the human role in space. *Enterprise,* the star of the day, began its role as the first icon for the new era. After serving its purpose as a test vehicle, *Enterprise* later would become NASA's marketing agent, appearing in the United States and abroad as the symbol of routine access to space.

Although NASA's press materials were rather restrained in their claims, media reports of the shuttle's debut in the general and aerospace press were highly favorable. Terms such as "start of a new era" (*New York Times*), "next exploring era" and "a new chapter in the history of flight" (*Baltimore Sun*), "a completely new breed of manned space vehicle" (*Chicago Tribune*), "a revolutionary new era in space" (*Newsweek*), and "a new space era" and "routine space transportation" (*Aviation Week & Space Technology*) indicated that the media had heard the main theme—a new era—and some of the subthemes—routine spaceflight for practical purposes. NASA used the debut of the shuttle to roll out a new ideology for the next era in human spaceflight, and the media cooperated by adopting those themes in both news reports and editorial commentary.

The new framework of meaning served to rejustify human spaceflight and recapture public interest. "A new era" in space transportation set human spaceflight into a long tradition of optimistic, progressive advances toward a brighter future. The cultural resonance of "a new age" or "new era" echoed a key concept of national identity—America as a new world. The frame of newness also harkened to a history of American innovation in mobility and transportation systems. Just as railroads, automobiles, and aircraft had brought about new eras in travel, with widespread impact on society and commerce, a new era in space transportation held similar promise. Placing human spaceflight and the shuttle into this historical frame—radically different from the pioneering race of the 1960s—gave it a solid and familiar foundation.[14]

NASA promoted the new-era ideology through varied channels, including informative, colorful public affairs brochures distributed to the media and elsewhere. As soon as the decision to develop the space shuttle was made in 1972, the agency began to craft the new-era message for public consumption. NASA commissioned space artist Robert McCall, who had worked on the 1968 film *2001: A Space Odyssey,* to paint scenes of typical shuttle missions for a color brochure that literally framed new ways of doing things in space. A simpler 1977 pamphlet titled *The Shuttle Era* claimed, "Now a new era nears . . . the coming of age in space" when people will be able to do important work there in ways never before possible. About the same time, Rockwell began to release public relations materials to promote "A Promising New Era."[15]

Routine space transportation was the central tenet of the new era. Spaceflight would no longer be a pioneering adventure; it would become commonplace and practical, in earth orbit, not outward-bound. In a burst of metaphors, NASA materials claimed that people would travel a highway to space in a workhorse that would operate like an airliner. Those mixed images might have been a clue that the new-era ideology of routine transportation still needed work on its coherence. The effort to make something practical sound exciting occasionally produced such rhetorical dissonance.

NASA further elaborated the concept of routine access to space with purposes that could appeal to special interests and make sense to the public at large. Commercial enterprise could use the shuttle to cash in on space by launching satellites or developing manufacturing capabilities there. Knowledge would increase as observatories were placed in orbit or scientists conducted laboratory research in space. Regular delivery of defense department payloads would enhance national security. Allegedly, all these activities on the shuttle would lead to practical benefits on earth. NASA thus plugged into the frame something to appeal to each necessary constituency—business, science, and military—and purposes that moreover would resonate with the public.

With promised economic, scientific, and security benefits, citizens could understand a practical approach to human spaceflight. Add to that the typical American consumer's desire for the latest-model vehicle or the newest technology, and "the new era of routine space transportation" was a potent frame for human spaceflight on the space shuttle. In this context, the purpose of human spaceflight was useful work. The shuttle served as a meaningful icon for this ideology. To see the broad-winged shape of the orbiter or the whole launch configuration with boosters and propellant tank stacked together was to recognize the new era—and new meaning—of human spaceflight.

Often called at that time "America's space truck" or a "workhorse," the shuttle better resembled another familiar transport vehicle for passengers and freight—an airplane the size of a Boeing 737 or a DC-9 jetliner. Yet both a freight truck and freight aircraft had the same basic purpose and configuration: a small cabin for people ahead of a large cargo container. Spaceplane and space truck became interchangeable terms for the shuttle that flew in orbit for the same purpose as an eighteen-wheeler tractor-trailer riding on the highways. As described in NASA's promotional materials and in the press, the space shuttle could be associated with familiar norms in the ordinary workday world—commuting to work, hauling cargo, scheduled departures and returns, economical travel, a shirtsleeve work environment, and commercial enterprise. Promotion of the shuttle hinted at the values of the middle class and the dignity of ordinary work, suggesting that spaceflight would no longer be a rare privilege. The space truck might democratize spaceflight and someday make space travel possible for anyone.

Initially, the primary emphasis in promoting the shuttle was on the innovative transportation system itself, not the enhanced value of onboard humans in the new era. People were oddly absent from these early graphic depictions, even though texts extolled the many activities that the shuttle would enable. The shuttle shape alone served as the icon for the virtues of the new vehicle and new era in spaceflight. The human element appeared later, when flights yielded imagery of astronauts living and working in space.

The *New York Times* director of science news, John Noble Wilford, was among the first journalists to introduce the shuttle-era imaginary to the public. His 1977 feature article, "Another Small Step for Man: Shuttling into Space," laid a bridge from the past to the future as *Enterprise* engaged in atmospheric flight tests. Echoing Neil Armstrong's famous words on the moon, Wilford placed the shuttle on the next rung of the ladder to humanity's destiny in space and greeted it as a revolution in space travel. He foresaw that the "era of the spaceplane" meant hauling orbital freight on regular flights and handling satellites for the three R's—release, retrieve, repair. He understood that the shuttle would not be used for exploration. But, because it would offer the ability to do new things in space, the shuttle might have as far-reaching impact as the automobile and airplane. At the end of the 1970s (just a bit prematurely), Wilford announced a variant of the new-era concept: the "Commuting Age Dawns in Space."[16]

When the new era truly dawned in 1981 as space shuttle *Columbia* roared into orbit on the first shuttle mission (STS-1), the new frame of reference crafted by NASA and presented in the media was in place. A different construction might have been possible, perhaps a mythic journey or another coherent metaphor, but none emerged. Already skeptics and critics weighed in, but the news media in near-unison trumpeted the new era of routine transportation to space. *TIME* magazine especially spotlighted it with two cover stories in early 1981. The first, in January, featured the pristine space shuttle on the launch pad beside the words "Aiming High." The main articles acknowledged both the practical—a round-trip truck for milk runs to the heavens—and the sublime—a reviver of old dreams—nature of the shuttle. In April, after the successful first mission, an exultant "Right On! Winging into a New Era" headline emblazoned the cover image of *Columbia*'s historic return. The shuttle-era human spaceflight imaginary was off to a fine start.[17]

The first shuttle mission received globally positive media coverage in recognition of the technologically revolutionary approach to spaceflight. It was actually an orbital test flight to check out the systems and verify that the shuttle orbiter, boosters, and tank all operated as planned. But it was an especially gutsy test flight because the total vehicle had not undergone any unpiloted test flights. After a spectacular launch, nominal checkout in orbit, and perfect descent and landing, NASA and the media proclaimed the quick two-day mission a great success. Main themes in headlines and news reports included national pride, hope for the future, and technological achievement—the beginning of a new space odyssey. Such rhetorical flourishes as "magnificent" and "breathtaking" (*TIME*), "renewed faith in technology" (*Newsweek*), "a mission of perils and hopes" (*New York Times*), "an astonishing technological triumph" (*New York Times*), "a reaffirmation of U.S. technological prowess" (*TIME*), and "a space odyssey" (*Chicago Tribune*) conveyed a shared sense of exhilaration at the resurgence of U.S. human spaceflight. When astronauts John Young and Robert Crippen returned from going to work in space, President Ronald Reagan remarked in greeting, "Through you, we feel as giants again."[18]

After the requisite jubilation, it was time to shift attention to the next missions and the promise of routine flights. Three more test flight missions occurred in 1981 and 1982 to further exercise the spacecraft systems. Each of these encountered some technical problems—an early return prompted by a fuel cell failure; a landing at White Sands, New Mexico, forced by weather; and loss of solid rocket boosters due to parachute failure. In addition, a noticeable number of protective heat shield tiles detached or suffered damage on each of the test flights. These issues were within the realm of what might be expected in test flights but hinted that achieving routine operations might take a while longer.

With the fifth mission in November 1982, the shuttle got down to business on its first designated operational flight. The first four-man crew ever launched into space went to work delivering two communications satellites for commercial customers. This was the first of twenty missions to deliver communications satellites for commercial and government customers, plus four missions to retrieve and return or redeploy satellites. Nine missions for the Department of Defense involved carrying mostly secret payloads into space and possibly leaving some there. Crews of twenty-six missions delivered scientific satellites, observatories, and solar system explorer craft into space. Most of these missions occurring in the 1980s and early 1990s confirmed a popular nickname for the shuttle—America's "space truck." Coincidentally, Americans' ownership of trucks escalated in the shuttle era, with retail sales tripling from 1980 to 1995 and trucks continuing to rise in popularity and sales for ten more years.[19] A truck for doing the work of spaceflight probably rang true to a pickup-truck-loving public.

Shuttle crews good-naturedly accepted the space truck label, adopting the motto "We Deliver" and making signs for "Ace Moving Company," "Ace Satellite Repair," and "Ace Observatory Delivery Co." to display on at least ten missions. Editorial cartoonists had a field day with the trucking concept and initially treated the shuttle with good humor. The *Washington Post* called it "The Space Truck" in a first launch day editorial.[20] Although NASA occasionally called the shuttle a space truck, there was some sensitivity about such a prosaic term for the remarkable flying machine and also about freight-hauling, a task done well by rockets, as a proper and costly role for human spaceflight. "Space truck" generally faded from use after the 1986 *Challenger* catastrophe, when the orbiter was destroyed and its seven crewmembers died while launching on a mission to deploy a communications satellite.

For the majority of its first twenty-five missions, and most of its 135 total missions, the shuttle indeed served as a space truck. Its large payload bay flew loaded with the cargo it was designed to carry. In accord with the human spaceflight rationale for the shuttle era, these missions accomplished productive work that could benefit life on earth. Although not yet achieving the hoped-for flight rate, the shuttle flew frequently enough that it began to seem routine. As the flight schedule increased from two missions in 1981 to nine in 1985, and again from two in 1988 to a steady seven and eight in 1992 through 1997, NASA logged successes that validated its rationale for human spaceflight.

How was the new era of spaceflight perceived by the public? A brief survey of reporting and editorializing about spaceflight during the first five years of the shuttle era indicates how the spaceflight ideology fared in practice. Commentaries on the first twenty-three shuttle missions (1981–85) in a major newspaper that followed the space program closely, the *New York Times,* are "reality checks" for assessing what actual spaceflight meant in the new era and are a reasonable gauge of public perception. Greeting the shuttle as a bold new approach to human spaceflight and the first mission as a triumphant return to space, the paper proclaimed "Columbia . . . Opening a New Era of Space Flight." Yet chief shuttle observer John Noble Wilford cautioned from the outset that the future was by no means certain; it might prove difficult to fulfill the optimistic predictions of the new era.[21]

A week before *Columbia*'s first launch (two years behind schedule), Wilford published a long, thoughtful essay on "Space and the American Vision." Four years had elapsed since his "Shuttling into Space" article—years during which the shuttle had been plagued with technical problems, cost increases, and delays. Wilford again explored the meaning of the new era of human spaceflight, but now the routine transportation scheme did not seem quite as plausible or resonant as before, and the shuttle had not even flown yet. There was a note of ambivalence about the shuttle era in his rhetoric as he tried to reconcile America's spacefaring destiny with the spaceplane's mundane mission of hauling orbital freight.[22]

Because thirteen of the first twenty-three missions were indeed freight-hauling flights to deliver satellites for commercial customers, *New York Times* reporters generally conveyed shuttle mission news within the routine transportation frame, featuring the three R's of space trucking (release, retrieve, repair). But they also made room in stories for questions about the cost of shuttle missions and reported all manner of technical glitches and delays that belied the ideology of routine spaceflight. The terms "failure," "delay," and "problem," repeated frequently in news accounts, subtly challenged the accepted frame of reference and sowed doubts about the fit between ideology and reality. Even so, the shuttle came to be understood as a space truck delivering large cargos to orbit, and successive satellite deliveries helped to establish a semblance of routine spaceflight.[23]

Attention to five missions in 1984 and 1985 elevated the space truck to new heights of interest by putting humans squarely in the focus. These missions added a Buck Rogers gloss to the notion of routine work in space and made the job of human spaceflight more intelligible.[24] The common theme of these servicing and salvage missions was satellites gone awry, humans to the rescue. The drama of astronauts flying away from the shuttle in jet backpacks, grappling errant satellites, wrestling them into the payload bay, and then conducting repairs put a human face into the new-era frame. The shuttle's image broadened from delivery truck to tow truck to service station, and the astronauts earned praise as orbital repairmen. Extravehicular activity (EVA) on these missions was a visually effective way to demonstrate human capabilities in space. The missions showed off new astronaut tools—the maneuvering backpack, remote manipulator system robotic arm, and power hand tools—so the idea of working in space became specific and concrete. Photographs and videos of these exploits enabled NASA and the media to convey through imagery the message that nothing quite like this had been done before.

By the end of 1985, with twenty-three shuttle missions completed, the *New York Times* and other news media had validated the new era of routine space transportation ideology as the meaning frame for human spaceflight. However, a noticeable current of critique ran through some of the news reports, and more so in editorials and opinions. Alert journalists noted that weather or technical problems delayed about two-thirds of the launches; several missions were delayed in returning or brought home early for the same reasons; and five years into the new era, the launch schedule was always subject to change. By these measures, "routine" transportation seemed elusive.

Of the satellites deployed from the shuttle, enough had failed to reach their intended orbits or operate properly that salvage missions were required, making the satellite deployment role for the shuttle look less rosy even though the astronauts displayed impressive retrieval and repair skills. Worrisome repeated problems such as damaged tiles, fluid leaks, computer malfunctions, locked brakes, and blown tires also clouded the picture of routine transportation. Occasional serious anomalies discovered after landing—evidence of a fire and explosion in the engine compartment, a large hole in a wing with partial melting of the structure—gave pause for observers to wonder how safe the shuttle really was. Despite the frequency and variety of missions in this new era, evidence mounted that human spaceflight was not yet routine.[25]

Only a few of the early shuttle missions provoked editorial commentary in the *New York Times,* which began to challenge the ideology of routine space transportation and useful human spaceflight. A skeptical editorial—"Is the Shuttle Worth Rooting For?"—appeared on the eve of the first shuttle launch. While acknowledging the shuttle as "an unquestionable technological achievement," the editors noted that it was "a technology in search of a mission" that might become a white elephant. The reason for their ambivalence: uncertainty that the shuttle would really cut the cost of operating in space. A few days later, the editors tempered their end-of-mission congratulations with the question, "Now that we own a successful space shuttle, what do we do with it?" Their standard: "What can a reasonable society afford?" The next editorial on the shuttle suggested limiting the number of spaceplanes to allow for continued planetary exploration.[26]

To mark the third successful shuttle mission, the *New York Times* acknowledged that *Columbia* "almost succeeded in placing the stamp of routine on shuttling into space," but charged that NASA was not using the magnificent machine with sufficient style. It deserved a purpose greater than trucking freight. In this instance, reality fit within the routine transportation frame, but the frame itself was challenged as unimaginative. However, the *Times* tendered no alternate frame.[27]

The tension between spacefaring and freight-hauling was a latent stress in the new-era ideology of routine transportation. Wilford's occasional reflections on the shuttle missions showed the stress fractures in this rationale and revealed how it was becoming dissonant, rather than more resonant, with some important cultural values. "This is no adventure in exploration; this is a freight run," he wrote upon witnessing the eighth launch. It did not inspire the same thrill nor satisfy the imagination as well as the missions to the moon.[28]

Wilford began to try to reframe human spaceflight by defining for it a purpose worthy of a spacefaring people with a tradition of exploration. With NASA under pressure to make spaceflight an economical business, he argued that the nation should aspire to a new vision of its future in space. Although the shuttle and a future space station would expand human activities in space, he looked to the robotic voyages of discovery in the solar system as the model for inspiring wonder and rekindling the spirit of the Apollo era. Despite NASA's efforts to sell the public on space utilization, Wilford and others began to long for the more romantic ideology of exploration.

Editorial cartoonists across the country also followed the shuttle and human spaceflight, and they helped make the space-shuttle-as-truck an icon (fig. 2.1). The metaphor was too ripe with comic potential and parody to resist. Editorial

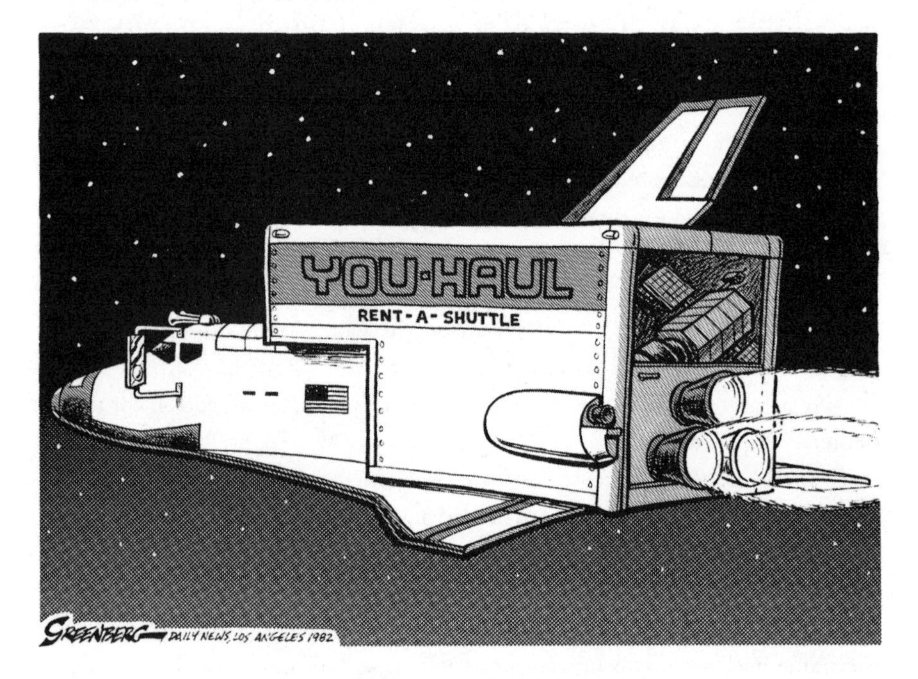

Fig. 2.1. Clever cartoonists transformed the shuttle spaceplane into a space truck in many guises—moving van, delivery truck, tow truck, service station, and others—to convey the practical missions it would serve. © Steve Greenberg for *The Daily News* of Los Angeles.

cartoons on the space shuttle offered a range of perspectives on the new era of spaceflight ideology, as they quite literally distilled an idea or opinion within an inked frame. Other members of the profession no doubt understood the motive expressed by one syndicated cartoonist who drew the shuttle: "I like turning symbols upside down and inside out . . . taking familiar symbols and restoring their meaning by looking at them with new eyes."[29]

In the early 1980s many cartoonists responded to the idea of routine space transportation with pride or humor. They tended to treat the first shuttle mission as a patriotic and technical triumph, using Uncle Sam and the U.S. flag to represent America's destiny in space. One depicted a shuttle timetable and boarding stairs to suggest routine flight (fig. 2.2). As flights continued, they drew the shuttle as a tow truck or service station with astronaut handymen repairing satellites. They depicted the foibles of launch delays and technical problems—a shuttle on the launch pad covered in cobwebs, suited astronauts growing old while waiting to fly, tiles falling off the shuttle, a tanker truck of superglue at the pad, a countdown clock with a ridiculously high number marking the time until launch.[30] Such cartoons were humorous but barbed, pricking the shuttle enterprise for its imperfections and challenging the status

Fig. 2.2. Proponents promised an era of routine spaceflight, but would there be enough customers for frequent, economical service? A 1981 Herblock Cartoon, © The Herb Block Foundation.

quo. Pulitzer Prize winner Pat Oliphant, whose repertoire included spaceflight, remarked, "For any cartoon, you've got to care about the issue, or you don't do it. And then you've got to find the way it should be done. . . . I'm looking for the magic combination of the message and the drawing just melding perfectly."[31]

Editorial cartoonists, inspired by the news and their own idiosyncratic perspectives on things, independently endowed the shuttle and human spaceflight with meaning inside the frames they drew.[32] Their charter for the shuttle, as for other topics, was to distill the essential meaning of things stripped of hype and pretense. Perhaps earlier than other observers, they began to see (and lampoon) a misalignment of NASA's ideology and reality.

It should come as no surprise that the Smithsonian National Air and Space Museum mounted an exhibit featuring the space shuttle, for the museum exists to present historic achievements in aviation and spaceflight. When it opened on the National Mall in 1976, the museum displayed a rich collection of space artifacts from the recently completed Apollo missions to the moon, as well as from Mercury, Gemini, Skylab, and planetary missions, plus towering rockets and missiles. Visitors encountered there under one roof the span of flight from balloons and the Wright Brothers to the heroic present. The museum soon became a pilgrimage destination for visitors to Washington, D.C., and a highly popular tourist attraction, one of the most visited museums in the world.

What is surprising is that the National Air and Space Museum staff began preparing a shuttle exhibit well before the new spacecraft flew and had *America's Space Truck: The Space Shuttle* in place by the first launch. Placed in the prime real estate of Space Hall, the large skylighted gallery on the main floor anchored by the "missile pit" and Skylab, the space shuttle exhibit remained on display under that title for a decade. Curators kept it under continuous revision to stay current with shuttle flight history as it happened.[33]

The initial purpose of the exhibit was to familiarize the public, millions of whom visit the museum each year, with the Space Transportation System in much the same ways as NASA and industry promotional materials, to inform people in advance about the revolutionary new vehicle and the new era of spaceflight. While this educational effort had been under way in trade show exhibits, published materials, and the media, such advance placement in a history museum was unusual, if not unprecedented.

The *America's Space Truck* exhibit script fairly well followed the pattern of NASA's promotional materials, introducing the components of the vehicle and its mission. Lacking flown artifacts yet, the exhibit featured a sixteen-foot-tall 1:15-scale model of the shuttle on its mobile launch pad and transporter, donated by Rockwell International. Models from earlier research programs and the shuttle design studies supported a historical overview of the quest for a reusable spacecraft. After explaining the basic elements of the shuttle and its operation, the exhibit asked the crucial question: What do you do with a space shuttle? The answer given: use it as a "freight carrier" to transport objects into space and eventually to support the construction of space stations.

In matters of space, and sometimes aviation, the National Air and Space Museum has a close relationship with NASA. The space agency is the source of space artifacts and occasional funding for exhibitions and public programs. When the museum opened, NASA Headquarters was housed directly across Independence Avenue, and the museum's director was Apollo 11 astronaut Michael Collins. Although the museum is an independent bureau of the Smithsonian Institution, it is often regarded as the de facto NASA museum. The *America's Space Truck* exhibit framed the space shuttle as positively as NASA had, as a vehicle for routine, economical human spaceflight. Further, the exhibit canonized the space shuttle in the nation's most hallowed place of honor for achievements in flight—even before the first attempt at liftoff.

As soon as the first shuttle mission ended, the museum began to update the exhibit with photographs and small artifacts to keep abreast of history in the making. The addition of garments worn by Sally Ride, Guion Bluford, and Senator Jake Garn marked historic "firsts"—the first American woman, African American, and elected official in space. The exhibit also documented operational firsts—satellite capture, satellite repair, laboratory research—with images and tools to depict the reality of working in space. Even after revising the exhibit to acknowledge the 1986 *Challenger* tragedy, the museum kept the title *America's Space Truck* as a persistent icon for the new era of routine spaceflight into the 1990s before changing it to *Space Shuttle.*

Despite mission successes, the shuttle program fought a losing battle to achieve the optimistic launch schedule and cost goals announced when the program began. Actual launch schedule history and shuttle operational costs were the toughest measures of the adequacy of the routine spaceflight rationale and thus the harshest basis for critique. The most strident critics of shuttle-era human spaceflight labeled it a losing cost-benefit operation. The business model did not yield more economical space transportation, and business-oriented counter-frames challenged the meaning of spaceflight.[34]

A corollary to the new era of routine space transportation promoted spaceflight as a business. NASA claimed that the reusable shuttle would lower the cost of spaceflight and make transportation to and from earth orbit economical.

The foundation for shuttle-era spaceflight would be a business model inspired by the commercial airline industry. NASA managers studied airline operations and sought to attract a customer market, contracted with payload owners for orbital flights, plotted the mission manifests for a fleet of orbiters, and calculated the operating margins to turn spaceflight into, if not a flourishing enterprise, at least a break-even business. With a sufficient number of vehicles and frequency of flights, the shuttle might bring down the cost of spaceflight enough to pay for itself.

NASA used this business model to defend the shuttle against critics who argued that the program was unnecessary and too expensive, and it dovetailed well with the concept of routine transportation for useful work in space. Transportation businesses on earth—interstate trucking, railroads, shipping, as well as airlines—were familiar analogues to give meaning to a space transportation enterprise. Blending these concepts enhanced the rationale for the shuttle by appropriating some elements of the adversaries' position, a sometimes useful strategy for broadening the appeal of an agenda. This model emphasized that human spaceflight meant efficiently running a customer-oriented business for practical benefits, if not profits.

The business-model frame proved vulnerable to critique by standard business accounting and auditing principles; it invited measurement of costs and gains. NASA had provided the quantifiable metrics for judging the performance of human spaceflight: flight rates and flight costs. As the shuttle became operational in the 1980s, it was not difficult for stakeholders in the business to do cost-benefit audits and assess the return on investment in human spaceflight. The value of work performed by the astronauts was more difficult to measure quantitatively, so the cost of operating the shuttle served as the primary measure for judging the value of human spaceflight. Thus, the business frame that was meant to promote the shuttle also became a frame for critiquing it.

Before twenty missions had flown, Wilford wrote a piece measuring actual performance against promise, in effect measuring the fit between the routine space transportation ideology and reality. Using such metrics as number of missions projected versus missions flown and number of satellites scheduled versus number orbited, he showed the large gap between expectations and reality. These discrepancies were prompting a reevaluation of the shuttle program by its customers and critics, and even its proponents. Despite mission successes, the framework for human spaceflight in the shuttle era was getting out of alignment with reality.[35]

Other observers also subjected shuttle-era human spaceflight to cost-benefit analysis and found that the numbers did not add up to economical space transportation. Several such critiques appeared in popular magazines in 1985, the shuttle's busiest year, with nine missions. Titles alone telegraphed the writers' attitudes: "It's Pay Off or Perish for the Shuttle," "Success amid the Snafus," and "The Shuttle: Triumph or Turkey?" The latter was one of the most strident in its appraisal of cost, technical failures, maintenance demands, uncertain schedule, deployment mishaps, and other shortcomings against the promises of routine space transportation. Historian Alex Roland wrote, "Judged on cost, the shuttle is a turkey. . . . It costs too much to fly. . . . And cost is the principal criterion by which it should be judged." He and others used the icon of human spaceflight to attack the credibility of NASA's new-era ideology for its unrealized promise of routine, reliable, economical space transportation.[36]

The shocking launch catastrophe of the twenty-fifth shuttle mission, the *Challenger* accident, crystallized the critique of this ideology and brought into high relief reservations about routine spaceflight. Well before the official investigation of the accident pointed to economic pressures and technical issues that put shuttle operations in unacknowledged jeopardy, journalists and public policy analysts focused their scrutiny on the shuttle's record. Realization dawned that the tragedy was waiting to happen, rooted in the design history of the vehicle, flawed decision-making in its operation, and pressures to achieve the promised flight rate. In other words, the ideology of routine spaceflight itself bore some of the blame for the avoidable loss of *Challenger* and its crew. This event was a blow to NASA's optimism about the shuttle's potential and a pivot point for sober reassessment of the shuttle's effectiveness.[37]

The catastrophe might have prompted the permanent grounding of the shuttle fleet and brought about an early end of the shuttle era due to concern that this kind of spaceflight was too dangerous. Had that happened, the U.S. human spaceflight program would have collapsed. Instead, flights were suspended until NASA resolved specified problems, and policy changes stripped the shuttle of some of its roles. Two years later, shuttle flights resumed. Military payloads and commercial satellites gradually migrated to other launch vehicles, and NASA increasingly used the shuttle for science and technology research in preparation for eventual space station construction.

During the years of developing and beginning to fly the space shuttle, NASA also sought to revitalize its image for the new era of space transportation. An organization's identity lodges not only in its words and actions but also in its "brand"—the reputation of its name and public awareness of its logo or product. NASA's brand had weakened with dwindling support for human spaceflight after the moon landing, and its mission became politically less urgent than that of other social programs. In this general malaise, NASA struggled to retain its stature and funding as an indispensable national resource. What better time to burnish its image than in the transition to a new program? High-level NASA staff launched an effort to update the agency's brand by creating distinctive symbols that would express or align with shuttle-era values. The agency sought a new look to match the new human spaceflight ideology, sometimes hiring professionals to assist and sometimes proceeding somewhat amateurishly on its own. Records of internal deliberations over logos and "product" names reveal how intently NASA worked on branding its identity in the shuttle era.

Under a federal graphics improvement program initiated by the National Endowment for the Arts in 1972, a design review panel examined NASA's graphic communications profile. They reported a lack of graphic standards and consistency in the agency's visual identity and found the NASA insignia or logo designed in 1959 to be a dated and unsatisfactory symbol. The dark blue circle with white letters and details bisected by a red airfoil, informally called the "meatball," was a busy design with so many elements that it was difficult and expensive to reproduce (fig. 2.3a). The reviewers reported that the traditional insignia and graphic practices did not well reflect "the most highly technological, exciting, and contemporary agency" that NASA aspired to be.[38]

NASA then hired a professional design firm to create a new logo and style guide based on design excellence. The firm responded with a unified visual communication system to manage the agency's graphic identity. Introduced in 1975, the primary element of identity was a stylized letterform logo of curved strokes resembling the letters N-A-S-A; it would appear in a single color, either bright red, black, white, or gray, depending on the background (fig. 2.3b). The

Figs. 2.3a and 2.3b. The squiggly NASA "worm" logo (*right*) replaced the original round "meatball" logo in 1975 as a more modern symbol for the space agency. The stylized logo branded the early shuttle era, but NASA revived the traditional logo in the early 1990s. NASA.

announced purpose of the streamlined new logo was practical: to "improve visual communications with the public" and "reduce design costs" by standardizing the graphic identity of the agency. Scornfully called the "worm" by partisans of the old "meatball," the new logo had a modern look and a symbolic meaning. According to NASA administrator James Fletcher, the logo reduced the letters N-A-S-A to their simplest form, "giving a feeling of unity, technological precision, thrust and orientation toward the future." He noted that the old graphic insignia had "detracted from the Agency's reputation as progressive and successful." Explaining the graphics change to the NASA workforce, Fletcher stated, "We at NASA believe that design excellence is not a luxury, but rather a necessity . . . to improve communications with the citizens of our country . . . so that design becomes a tool in achieving the program objectives of NASA." The new *Guide to NASA Graphic Standards* further acknowledged the vital interplay between the agency's mission and its graphic identity.[39]

This logo became a coherent element in NASA's strategy of framing a new era of spaceflight. Just as the space shuttle was a revolutionary spacecraft, the NASA logo was also innovative, and it too aimed for economy. The marketing package for selling the public on spaceflight now included both technical and symbolic elements. The new "NASA worm" soon supplanted the old "meatball" on everything, from identification badges and stationery, to building signs, to agency vehicles on the ground and spacecraft headed for launch, including the shuttle. Despite some institutional grumbling and resistance, everything associated with NASA would bear the same bold new logo. Sheer ubiquity would guarantee its becoming the signature of the agency, the icon that immediately would signify NASA in the public eye.

A new logo for a new era, a modern logo that suggested a thrust toward the future: the timing was perfect. The new logo was "launched" in time for the U.S. bicentennial celebration and also for *Enterprise* to appear in 1976 as the first spacecraft painted in the new livery with the NASA "worm" boldly on its body and wings. A decade later in 1984, this sleek logo and companion design standards manual, along with the "Going to Work in Space" poster series, received Presidential Design Awards for excellence in visual communications in the first such federal competition. NASA's new logo prevailed for fifteen years, until an administrator with a different agenda banished it in 1992 and restored the sentimental old logo.

After settling on a new brand to convey the agency's new-era identity, NASA faced another identity issue: what to name the space shuttle vehicles. The eventual announcement that the orbiters would be named after sea vessels used in exploration stirred no controversy. In fact, the process of deciding the orbiters' names was somewhat fraught, because it brought to the fore the main issue: how to characterize the shuttle era of spaceflight. The orbiter vehicles would be the visible icon of the new human spaceflight ideology, and so it would be sensible for their names to fit coherently within that rationale. Having been preempted in the naming of *Enterprise,* NASA was determined to retake the initiative for naming the flight orbiters.[40] What would follow the Mercury, Gemini, and Apollo tradition? More mythology? Names from American history or geography? Scientific names?

Over the course of nine months in 1978, two separate committees at NASA Headquarters labored to deliver a worthy set of names for the four spacecraft to come. One committee of space shuttle and public affairs personnel began with a list of about a hundred random names suggested by NASA, contractor personnel, and the public before *Enterprise* was named. They decided to recommend "names having significant relationship to the heritage of the United States or to the shuttle's mission of exploration" and forwarded a ranked list of fifteen names to higher management for comment. Heritage names like Constitution, Independence, and Liberty certainly fit within the spaceflight framework. Names related to exploration were somewhat problematic, because the shuttle's mission was decidedly practical and utilitarian.[41]

Because several of the names were those of illustrious warships, the list returned to the committee for further work. They began to focus on new categories of names: aerospace pioneers, scientists, explorers, transportation vehicles, and astronomical phenomena. Unfortunately, few of the names in these categories seemed suitable for a high-tech spacecraft or its mission. The group considered Wright Brothers and Wernher von Braun, Copernicus and Galileo, Marco Polo and Lewis & Clark, Conestoga and Sternwheeler, Quasar and Pulsar, and many more names for the most sophisticated flying machine ever. They sent a list with those and similar names per category up the chain of command; at least one official sensibly found them all unsatisfactory.[42]

A more intramural committee formed to make further suggestions. This group also worked with thematic categories but refined them to explorers' vessels, American tradition, and stars or constellations. The committee's top preference was ships of exploration, and the names they forwarded for approval were those of British explorer Captain James Cook's vessels: *Endeavor* [*sic*], *Discovery, Resolution,* and *Adventure.* Yet, when the press release announced "Shuttle Orbiters Named After Sea Vessels" in early 1979, the chosen names were *Columbia, Challenger, Discovery,* and *Atlantis.* President Jimmy Carter's science adviser, allegedly after consultation with someone at the Smithsonian, had substituted the names of two U.S. research vessels and a different British ship to shift the balance toward American exploration history.[43]

NASA accomplished two objectives in naming the space shuttle orbiters. By deliberately choosing the names of ships of exploration, NASA set up a defensible intellectual rationale, albeit one that bordered on sleight of hand. It co-branded the new era of spaceflight as exploration, even though the mission of the shuttle was utilitarian, not exploratory. (One set of suggested names more true to the purpose of the shuttle did not make the cut: Progress, Benefit, Harvest, Advantage.) By spreading the aura of exploration over an endeavor intended to be routine and practical, NASA could appeal to the public's sense of spaceflight as adventure. NASA also gained control over its fleet identity. By naming all four orbiters as a group, it guaranteed a consistent image and avoided a public incursion like the *Star Trek* maneuver in naming *Enterprise*.[44]

The orbiter names thus became a conscious part of the framing of human spaceflight in the shuttle era and also gained iconic status in public awareness of spaceflight. The media aided this identification by using orbiter names as a more intelligible shorthand than the flight numbers NASA assigned to missions (STS 41-D, for example). Reporters called the flights "the *Columbia* mission," or "the *Discovery* mission," and the STS 51-L tragedy instantly became "the *Challenger* accident." Orbiter names became hallmarks for human spaceflight in the space shuttle era.

NASA also developed an official identification symbol for the Space Transportation System (STS) program. Depicting a stylized orbiter in upright launch position atop the external tank and solid rocket boosters over the words "space shuttle," the triangular design was meant to be a patch worn on flight suits and uniforms (fig. 2.4). NASA's public affairs office encouraged widespread use of

Fig. 2.4. This symbol of the space shuttle program, like the triangular delta mathematical symbol, telegraphed the idea of change. NASA.

the insignia "to promote program identity," and so it spread through the shuttle workforce on shirts, caps, hardhats, and office doors. This emblem also made its way onto souvenir items and toys in the commercial marketplace.[45]

The shape of the space shuttle stacked for launch and that of the orbiter alone abstracted easily into a triangle. Several aspects of shuttle technology shared the triangular shape—the triad of main engine nozzles, the orbiter's delta wings, the nose of a solid rocket booster—and the upward-pointed design evoked the idea of launch. Among engineers and mathematicians, the triangle is used as a symbol for "delta" or "change," as in "delta v" for change in velocity, written in an equation as "Δ." Subliminally if not intentionally, the triangle's meaning as a symbol of change made it an appropriate design element for the novel space transportation system.

This space shuttle patch could be augmented with further project, mission, or payload identifiers by adding a name bar or bars at the bottom. There was some concern that this practice would nullify the tradition of distinct crew patches for individual missions. Chris Kraft, director of flight operations at the Johnson Space Center and an advocate for workforce morale, urged that the crew patch tradition be continued, not only for its in-house use to acknowledge with pride one's work on different missions but also to serve the public. He observed that "the crew patch has also been an item of considerable interest to the public and has allowed the agency to maintain a continuing public interest in the flights," and he noted that the most popular items sold in the NASA shops were crew patches or items adorned with patches. Kraft recognized that spaceflight could be commodified into desirable articles for sale and that it was in NASA's interest to provide ways for people to connect with spaceflight through purchasing and owning such collectible items. His opinion helped save the crew patch tradition; it continued in peaceful coexistence with the new STS space shuttle patch, which never was modified with bars for specific missions.[46]

Both the STS space shuttle patch and toys in the shape of the shuttle soon appeared in stores. Shuttle toys spanned the age range from tub toys for toddlers to highly detailed models for mature hobbyists (plate 3). With wings and wheels, the orbiter was easily adaptable into a pull-toy on a string, a push-toy for little hands, a Hot Wheels or Matchbox vehicle for school-age kids, and a remote-control flyer for adolescents and older. The shuttle shape appeared as cereal bowls and dinner plates, drinking cups, puzzles, board games, swim floats, plush toys, a host of other play products, and inevitably on T-shirts, ball caps, sweatshirts, and jackets. The basically triangular shape of the shuttle orbiter proved remarkably resilient. It could be made short and stubby or long and slender, obviously not the proportions of the actual vehicle but still recognizably a shuttle.

Almost every major manufacturer of toys and models released shuttles in some form, and in packaging that often bore the curvy NASA logo or the triangular space shuttle program emblem. Brio, Fisher-Price, Hasbro, Hoyle, Kenner, Lego, Mattel, Monogram, Revell, Tomy, Tonka, and others targeted many of these products to children too young to be well aware of actual spaceflight. Some made the shuttle cute by turning its windows into eyes to make a friendly face and changing its proportions for little hands to grasp. Some shuttle toys depicted payloads or included astronaut figures. Manufacturers and merchants saw the potential of the space shuttle for fantasy play among young consumers and educational play among older children. Model kits typically were labeled as highly realistic and included fact sheets about the shuttle. Some package blurbs suggested that playing with these toys would make children smarter. A number of products bore the "seal of approval" of the Young Astronauts Club, a national organization whose mission was to stimulate young people's interest in math, science, and space exploration.[47]

The proliferation of shuttle toys in stores and homes alongside toy cars and airplanes was consistent with the overarching goal of the space shuttle era: to make routine spaceflight a normal part of American life. What could be more normal and routine for the next generation of space explorers than growing up with a shuttle toy? Indeed, if spaceflight toys, games, and models were part of their youth, the next generation might gravitate toward supporting or working in the space program, much as the first generation of aerospace workers did after devouring space fiction in their youth. It was in NASA's interest for these products to be "consumed" by children (and the adults who paid for them) to keep the public avidly connected to human spaceflight. As a toy, the space shuttle served to make spaceflight a commodity that could be consumed by children from the cradle on.

Another commercial success appeared in print, a book titled *The Space Shuttle Operator's Manual,* written by two educators working at the National Air and Space Museum. First published in 1982 and revised in 1988, both editions remain for sale and probably can be found in the home or office of most spaceflight aficionados. The book was an indispensable popular reference that addressed the reader as an astronaut preparing for a flight. It explained the features of the vehicle and the phases of a typical mission, and how to do things—make a meal or use the toilet, sleep and dress, use the orbiter's robotic arm or do a spacewalk, deal with emergencies, and a host of other information to satisfy the curious. Foldout pages displayed the full layout of switches and buttons on the flight deck panels so the armchair astronaut could mimic some of the crew's procedures. Full of information but not technical jargon, this handy reference manual, like the toys and model kits, made the space shuttle familiar and offered individuals a share of the excitement of spaceflight. Although NASA did not produce the book, its popularity furthered the agency's public awareness goal.[48]

As NASA and the media framed the new era of routine spaceflight as going to work in space, they did not initially concentrate on the human element. Until shuttle flights began, promotional brochures and feature articles focused on the vehicle and its missions. Those writing about it almost endowed the shuttle with free agency by describing it as carrying out the useful tasks of delivering and retrieving satellites, carrying out scientific research, and one day supporting space station construction. It was almost possible to misunderstand the space shuttle to be an unpiloted craft.

Obviously, astronauts would be crucial to making routine spaceflight a reality and accomplishing the ambitious workload planned for the shuttle era. The astronaut corps already had a distinct identity for the prestige-driven space race, but the ranks had thinned and those still at work were less visible in public during the spaceflight hiatus of the 1970s. The gap between the debut of the shuttle in 1976 and the first launch into space stretched to five years as the final assembly and testing of *Columbia* ran into problems. During that period of delays, an event in 1978 drew attention to the going-to-work-in-space ideology of the shuttle era: selection of the first group of new astronauts. In another sign of new times, this generation included six women, three African American men, and a man of Asian descent. Although veteran astronauts would fly the first shuttle missions, this large, diverse group of fresh faces signaled that the shuttle program was ramping up for the new era of living and working in space. NASA began to expand and reconfigure the astronaut corps for the purpose of the new era of human spaceflight. When shuttle missions got under way, the public finally began to see how the role of astronaut and the face of the astronaut corps were changing to match the new spaceflight ideology.

Chapter 3 Astronauts: Reinventing the Right Stuff

NASA staff and news reporters filled the auditorium at agency headquarters in Washington, D.C., on January 16, 1978, for the announcement that a new group of astronaut candidates had been selected, the first called for duty in the space shuttle era. About a third of the first-generation astronauts remained ready to fly, but the goal of routine flight required a larger corps. NASA's first recruitment in ten years drew more than eight thousand applications. The agency selected thirty-five people to train for going to work in space, choosing for the first time women and a multiracial group. More than half of them would be mission specialists, not pilots—a new role for scientists and engineers in space.

During the press conference, reporters repeatedly asked about qualifications and selection criteria, hinting that NASA might have reduced standards for admission to the astronaut corps to meet gender and race quotas. The media and public construed astronauts as an elite group of mostly clean-cut white male military combat and test pilots. Now, a group of women and men, white and brown, many with Ph.D. and M.D. degrees, stood on the threshold of the shuttle era. The new candidates were not present that day; in fact, they had been notified of their selection only hours before. Two weeks later they would convene to meet the press. In their absence, and then in their presence, the media struggled to fit these newcomers into prevailing conceptions about the role and image of astronauts.[1]

For NASA, these two occasions rolled out its deliberate effort to reshape the astronaut corps. The multipurpose space shuttle vehicle was only one component of the new era in space; equally germane to the concept of routine spaceflight were the people who would fly and work in it. NASA took the lead in reframing the astronaut role, and the media interpreted these new astronauts for the public. But seeing new faces in space did not happen swiftly or without shedding familiar ideas. In the dawning new era, the shuttle-era imaginary for astronauts took shape as new images emerged to dislodge the old.[2]

From America's first seven space pilots to the latest crews on the International Space Station, astronauts are the human face of the vast spaceflight enterprise. Astronauts are the essence of NASA's brand; they symbolize competence and courage, both personal and institutional. Astronauts are a symbol of national technical excellence and leadership, success, and patriotism. They stand for pioneering, facing risks, and defying death in service to American ideals. Globally, they are the iconic embodiment of the spacefaring dream. People around the world generally view them as celebrities and heroes simply because they *are* astronauts. Even without knowing their names or missions, people are eager to pose for photos with them, shake their hands or seek their autographs, and bask in their aura for a memorable moment.[3]

Apart from their symbolism in the moment, astronauts also stand for a timeless cultural myth: the hero's journey. The astronaut as heroic figure is a social expectation inspired by the mythical archetype, science fiction, NASA and the media, the real risks of spaceflight, and other factors. Americans have readily framed astronauts as heroes, fitting them with a mythic ideology of virtue, valor, and excellence. These heroic expectations flowed through the pioneering years of spaceflight; would they likewise inform the working shuttle era?[4]

NASA initially set criteria for astronauts that very few could meet. The first thirty astronauts chosen in three groups from 1959 to 1963 to launch the U.S. presence in space were primarily test pilots, all but four of them drawn directly from the military. Those who were experienced in the risks and rigors of experimental high-speed, high-altitude flight offered the best chance of success. They understood aerodynamics and the disciplines of flight from pushing the limits of aircraft performance, so they could step into the astronaut job already proficient. They could also participate actively in spacecraft design, testing, and procedures development to make substantive contributions to the engineering process. Acculturated through military service to the nation, they were well suited to achieve America's early aspirations in space.[5]

Already decorated military heroes, the early astronauts also suited NASA's willingness to cultivate its own image. Realizing that the astronauts could connect the public with the space program and attract public support for the difficult challenges ahead, NASA carefully shaped a public relations package around them. NASA and the media presented the astronaut corps of the 1960s as the archetype of masculine heroic ideals: young, handsome, physically fit, intelligent, courageous, competent, tough-minded, fearless "all-American boys" and family men. That some astronauts were also brash was largely kept from public view. Astronauts were groomed to be the symbol of a nascent space agency on an urgent mission. Yes, they would carry out perilous flights into space; more important, they would also symbolize the best of America. The public soon idolized these idealized men.[6]

Except for two unsuccessful attempts to bring an African American pilot into the early astronaut corps, NASA's practice of drawing astronauts primarily from the very small pool of the best military test pilots effectively created an astronaut corps exclusively made up of white men. No women could meet the tight selection criteria, because none were flying high-performance military jets or serving as experimental test pilots. The U.S. military services did not then permit women to serve as pilots, so women had no access to test pilot schools. Although a group of female civilian pilots, privately evaluated outside NASA in 1960–61, passed the same battery of physical and mental tests administered to male astronaut applicants, they were not allowed to go through flight evaluations because the agency then had no interest in admitting women into the astronaut corps.[7]

From the silver-suited official group portrait released in 1959 to a contract with *LIFE* magazine for feature stories focused on the original astronauts and their families, the public greeted astronauts with awe and pride as the anointed heroes of the space age. Magazine covers, ticker-tape parades, and presidential welcomes at the White House signified their celebrity status. Years later, journalist Tom Wolfe scratched through the patina of hero worship to reveal machismo and rowdiness within the astronaut corps. *The Right Stuff* was not only the title of his award-winning book about the test pilot culture that produced the original astronauts; the phrase also codified in shorthand those qualities of excellence and bravado that characterized the astronauts of the space-race era.[8]

Through 1969, NASA recruited more pilot astronauts, inviting civilian test pilots to apply and selecting a number of them for duty. Military pilots remained dominant with the transfer of a group from the Air Force Manned

Orbiting Laboratory Program. Twice in the 1960s NASA recruited scientists who were screened and recommended by the National Academy of Sciences, in response to pressure from the scientific community to add researchers to the astronaut corps. Of the seventeen scientists (all men) who were chosen, one went to the moon on the final Apollo mission and three flew on the Skylab missions of 1973–74 before U.S. human spaceflight went into a lull until the space shuttle was ready to fly. These scientists never achieved the fame of the pilots or equal influence within the astronaut corps, but they were precursors of the scientist-astronauts who would flourish in the shuttle era.[9]

Three *TIME* magazine covers appearing within a few months of each other in 1968 and 1969 conveyed multiple meanings of the astronaut as icon at the climax of the space race. A December 1968 issue featured two flight-suited, faceless figures racing stride for stride to the moon, identical except for the color of their suits and the United States flag or Soviet Union's red star on their shoulders. The figures were symbols of the cultural values signified by those two emblems. In contrast, a month later the "Men of the Year" cover featured the three Apollo 8 astronauts just returned from the first human voyage around the moon. Their actual portraits and names appeared, but no spacesuits or helmets or spacecraft, nor a U.S. flag. Only a luminous full moon behind them symbolized the reason for honoring these individuals. Each face showed character in a calm, steady gaze with just a hint of a smile, projecting the maturity and confidence befitting bona-fide national heroes. In July, the magazine cover art presented a quintessential astronaut—standing on the moon and holding a large American flag. It represented Neil Armstrong, unnamed but recognizable in profile. Wearing a smaller flag on his shoulder, he became the iconic symbol of the United States and its triumph.[10]

This powerful ideology and iconography of the astronaut as national hero and symbol awaited the new astronauts serving in a less politically charged time and with a much different role to play. Would these precedents apply as well to them? How would the idea of practical, routine spaceflight be realized in imagery? What would become the icon for the new era? Hints began to emerge in NASA's planning for the shuttle and an expanded astronaut corps.

Having decided to pursue routine human spaceflight with the space shuttle, the United States would need a new kind of astronaut with scientific and engineering skills to make space a home and workplace. Furthermore, social changes in the 1960s and 1970s raised expectations for a more egalitarian astronaut corps. Under pressure for inadequate hiring of women and minorities, NASA committed to transforming the astronaut corps into a more diverse representation of America's finest. Astronauts would be a clearly visible marker of NASA's progress toward equal opportunity in gender and race.[11]

By January 1972, when President Nixon announced the decision to develop the space shuttle, it was clear that change was coming. He noted that the less physically stressful ride offered by the space shuttle would mean that "men *and women* [emphasis added] who have work to do in space can 'commute' aloft" without having to endure as much training as in the past. NASA administrator James Fletcher echoed this statement in his announcement, and it became a staple in NASA's space shuttle promotional materials. Although there were no women on deck yet, presidential and space agency rhetoric signaled that the shuttle would open the way into space for them and others besides test pilots.[12]

Of the two primary reasons for broadening the astronaut corps, the nature of the complex spacecraft itself demanded change. The shuttle needed pilots, but even more it needed engineers to operate its systems and technical payloads, and it needed scientists to conduct onboard experiments in a variety of research disciplines. A one- to two-week mission typically would require a crew of five to seven astronauts, only two of whom would be responsible for actually piloting the vehicle. The operational requirements for a shuttle mission bespoke a rebalanced astronaut corps in which scientists and engineers, called mission specialists, outnumbered pilots. With guidance from the scientific community and scientist-astronauts from prior programs, in the mid-1970s NASA began to define the mission specialist roles and expectations for the space shuttle era. In addition, the anticipated frequency of flights meant that well-trained crews must always be available; the existing astronaut corps was simply not large enough or versatile enough to meet the anticipated needs.[13]

The second reason was social equity. The time had come for the astronaut corps to look more like the U.S. population—that is, to include women, African Americans, and members of other minority groups. The Civil Rights Act of 1964 was opening new opportunities in education and the workplace. The policy of "affirmative action" in place since 1965 (modified in 1967) required the government to hire without regard to race, religion, color, sex, and national origin, and the Equal Employment Opportunity Act of 1972 barred discrimination based on any such categories. The most competent person who applied had to be hired. Increasingly sensitive to perceptions that it was an elite but closed organization, NASA needed to attract highly qualified candidates from a broader population. Without diminished regard for the earlier generation of astronaut heroes, the astronaut corps could no longer be a white male preserve.[14]

The shuttle vehicle and the goal of a more diverse astronaut corps were mutually compatible. As a larger, more complex spacecraft than the previous capsules, the shuttle could accommodate a larger crew to carry out different responsibilities. It was spacious enough for a modicum of personal privacy for mixed-sex crews. A versatile spacecraft capable of varied missions required crews with many skills, and that variety of skills allowed a much broader population, including women and minorities, to qualify as astronaut candidates. Given the astronauts' role as the face of the space program and the public's positive perception of astronauts as the "best and brightest," NASA hoped that public support might well increase if more people, especially youth, saw themselves reflected in a diverse astronaut corps.

As NASA strove to match its operational needs with the social pressure to diversify, it reshaped the idea and image of the astronaut to match the mission of the space shuttle. Instead of heroes, astronauts would be workers, highly competent professionals from a variety of technical fields. Some would still be military pilots, but many would be civilians from industry and universities. NASA reframed the astronauts as skilled engineers who would deliver and repair equipment or assemble a space station in orbit, or as scientists who would conduct research in space to deliver benefits for needs on earth. Eventually these astronauts became standard-bearers for the importance of science and technology education. The astronaut icon assumed these new meanings, yet retained some of the old. Recruitment of the first and subsequent classes of shuttle-era astronauts exemplified the deliberate shaping of message and reality to reinvent the astronaut imaginary for a new era.

Under the leadership of James Fletcher and his deputy, George Low, NASA spent several years refining its plan to encourage women and members of minority groups to apply to join the shuttle-era astronaut corps. As early as 1972, the astronaut office began to specify the technical criteria for pilots and the new mission specialist position responsible for spacecraft systems and payloads. Pilot qualifications were essentially the same as before, but those for mission specialists were reminiscent of those for the scientist-astronauts recruited in the mid-1960s. The primary requirement was a degree in the physical sciences, life sciences, or engineering. It was desirable but not necessary to be a pilot, and the physical standards for vision, hearing, and height were slightly relaxed. Otherwise, like pilot applicants, mission specialist applicants were subject to rigorous physical, medical, psychological, and interview requirements.[15]

The offices of public affairs and equal opportunity began to devise a recruitment strategy. NASA's challenge was to get the word out to ensure that not only typical applicants would emerge but also women and people of color would apply in sufficient numbers to compete well in the screening and selection process. Astronaut recruitment for the shuttle era strove to reach the broadest possible applicant pool to find the most qualified candidates and to demonstrate that NASA was serious about social equity and equal opportunity.

To address concerns and a lack of data about women's strength and physiology, NASA conducted several studies before and during the recruitment period. Volunteer female employees at the Johnson Space Center underwent treadmill tests and sessions in a lower-body negative pressure device to gain baseline data on female tolerance to cardiovascular stress. Other groups of women underwent isolation testing, bed rest studies of metabolism and biochemistry, hormonal and gynecological data collection, underwater neutral buoyancy trials, and simulated science missions so researchers could learn more about female physical and psychological responses to spaceflight conditions. These studies indicated that women's physiological adjustments and tolerance to simulated weightlessness were similar to those of men, and nothing emerged as a problem for women in space. At the same time, the original scientist-astronauts participated in studies to understand how to conduct experiments in weightlessness and better define the mission specialist role.[16]

In 1976 NASA launched an aggressive one-year recruitment campaign targeting women and minority group members through professional organizations, university science and engineering departments and graduate schools, college placement offices, government agencies, aerospace corporations, and other channels. The agency brought popular black actress Nichelle Nichols, recognized as Lieutenant Uhura of *Star Trek* fame, under contract to make public appearances and public service announcements encouraging applicants to the new astronaut corps. The call for applications explicitly stated that "minority and women candidates are encouraged to apply," a message that the news media repeated in headlines and stories.[17]

Thousands of applications arrived for NASA's consideration, but it was hard to know whether the recruitment strategy was succeeding. Applicants' names usually indicated gender, but there was little to indicate who might be members of minority groups, only clues such as mention of a historically black college or Hispanic locale on a résumé or a recognizably ethnic name. NASA's Astronaut Selection Board rated applications on their merits, and only when finalists arrived for interviews might their racial or ethnic identity become evident. For the first time, the selection board included a woman and an African American to screen, winnow, interview, and select the new astronauts.[18]

NASA manager George Abbey, perennial chair of the Astronaut Selection Board and for years the person responsible for assigning mission crews, carefully maintained the mystery of the selection process. He would say only that NASA sought self-reliant experts with initiative who also were good team players; it was not a job for egotists. Besides technical competence, NASA sought people who engaged in physically challenging experiences—high-level competitive sports, SCUBA diving or skydiving, mountaineering—or such demanding pursuits as music performance. The board also looked for evidence of both expertise and the ability to master new knowledge quickly.[19]

The first shuttle-era recruitment produced a class of thirty-five astronaut candidates (called Group 8) from more than eight thousand applicants: fifteen pilots and twenty mission specialists, six of them women and four minority-group men (three African Americans and one American of Japanese descent). More than 1,100 women applied. Of the twenty scientist-engineer mission specialists, six had been military pilots and another six held commercial pilot licenses. The media duly reported that this group broke the sex and race barriers that had existed since the first astronaut selection in 1959 but rarely noted why women and minorities had not previously qualified. NASA stressed that these were "the most competent, talented and experienced people available to us today" to avoid an impression that anything but ability had factored into the selection.[20]

In the announcement press conference, journalists grilled Johnson Space Center director Christopher Kraft about the selection criteria. Had NASA attempted to fill quotas or reduced standards in order to add women and minorities? Had any positions been lost to men in order to add more women to the group? NASA administrator Robert Frosch indicated that the affirmative action effort to find astronauts from a wider spectrum of American society had not compromised standards or selection on merit. All were highly qualified and motivated. One novice attracted special attention: Shannon Lucid, a biochemist and married mother of three, and at age thirty-five the oldest of the women. Journalists were curious about how motherhood factored into her selection or might limit her future assignment to fly. Although twenty-five of the male candidates were fathers of fifty children in all, no one raised the same question about them. Kraft answered that parental status did not matter.[21]

One reporter noted with prescience that this group promised "to become America's first true career astronauts. They are the men and women who will someday take the shuttle into space so frequently that its comings and goings likely will be buried in the back of the newspaper, if reported at all." Routine spaceflight meant that the astronauts might fly often but also become interchangeable and anonymous, humanized rather than idealized. The new astronauts likewise had a sense that fame as heroes would not be their lot. "By the time I get into space, the flights will have become commonplace," Robert "Hoot" Gibson said. "We will be just operators."[22]

The media welcomed the "new breed" of astronauts but tended to focus more on the novel presence of women than the racial integration of the astronaut corps. The *New York Times*, for example, ran a front-page photo of the women only. Some reporters made only passing reference to the African and Asian Americans, or ignored them altogether, but probed how women would make a difference in spaceflight, working in close quarters with men. Evidently untouched by 1970s feminism, they asked blatantly sexist questions based on stale notions of women's interests, roles, and appearance. Struggling to understand how women would fit into spaceflight, they lapsed into clichés about housekeeping, staying fit, and being well groomed in space. Answering for themselves, the women astronauts stressed professionalism. "We'll all have jobs to perform," Lucid said.[23]

Within days, editorial cartoons and comics began to play with the same stereotypes in framing the arrival of women in a heretofore masculine occupation. They drew the shuttle with frilly curtains—"The Woman's Touch"—or made jokes about lipstick and housekeeping chores in space. Political cartoonist Pat Oliphant drew a boarding crew of male astronauts giving their beverage orders to the female astronaut, identifiable by her long hair and figure-hugging spacesuit. Some comics expressed anxiety about possible in-flight infidelity, with a wife tagging along or exacting a promise of good behavior. Standard jokes about women asking for directions or backseat driving migrated into spaceflight sketches. Five years later, for the flight of America's first female astronaut, little had changed; one of several sexist cartoons depicted the shuttle with a clothesline dangling women's lingerie over the payload bay. If cartoons tap the pulse of social values, these suggested the mismatch between traditional perceptions of astronauts and the new version that NASA sought to create. Like journalists lobbing sexist questions to the first women astronauts but not to the men, cartoonists also seemed at a loss about how to adapt the icon to changing times.[24]

Immediately after the first shuttle-era astronauts were introduced in Washington, they traveled to the Johnson Space Center in Houston for a quick familiarization. It would be several more months before they would report for duty, but NASA gained some further publicity by this advance visit to their new employer. Among the photo ops staged during that visit, three particularly resonated with the message that it was a new era in spaceflight.

In one image, the six women candidates, in business attire, stood beside a space shuttle model the size of a conference table. They were smiling, with their hands resting at ease on the wings, payload bay doors, and nose of the orbiter as if they were claiming possession. The American flag and "USA" adorned the vehicle's wings in the foreground, and the women framed the V-shape of the spacecraft. In a similar group photo, they wore blue flight suits and posed with a white spacewalking suit as if they were ready to fly. The other image showed three males dressed in white pressure suits like those worn on the moon (the shuttle suits were not yet available). Soft "Snoopy caps" covered their hair and gloves hid their hands, but it was easy to see that two of the faces were decidedly dark and the third was a deep tan. These three men, the first African Americans selected to become astronauts, smiled confidently. Everything about the images telegraphed that these people were American astronauts.

These photographs were among the first to visually substantiate the idea of diversity in the astronaut corps. They showed far better than a list of the selectees' names in a press release that space had become an equal-opportunity workplace. Such images launched the iconography of the shuttle-era astronaut even before the newly chosen men and women started working at NASA. Later recruitments continued the trend toward diversity, although no other astronaut class had such a pronounced gender and racial mix. The standard group portraits of mission crews through the years vividly recorded the new norm (fig. 3.1).

Soon the new astronaut candidates moved to Houston and settled into intensive training in classrooms, labs, and specialized facilities. Without an unseemly sense of self-importance, they called themselves the TFNGs, the Thirty-Five New Guys. Their curriculum included all spacecraft systems and spaceflight operations, plus related activities such as navigation and communication, survival training, and underwater weightlessness training. Men and women candidates trained together and were expected to perform at the same level. This training put all the astronauts, with their diverse backgrounds, on the same footing and also gave NASA further opportunity to evaluate their performance before committing them to flight. The astronauts likened it to earning a master's degree in spaceflight.[25]

Group 8, class of 1978, made excellent progress and completed their introductory training ahead of schedule. They graduated from candidate to astronaut status and received technical assignments in various offices around the Johnson Space Center to acquire more skills in preparation for spaceflight. These newcomers earned the respect of the "old guard." Their training director, Apollo and Skylab astronaut Alan Bean, and John Young, the dean of the astronaut corps and first shuttle commander, admitted to being impressed.[26]

Media response to the "new breed" of astronauts was admiring but not worshipful. Journalists focused on the theme of diversity, pointing out that the new astronaut corps represented more varied backgrounds and abilities than before.

Fig. 3.1. One of the most diverse crews of the shuttle era, assigned to the ill-fated final *Challenger* mission, represented how the astronaut corps demographic had changed. NASA (S85-44253).

They also took a rather workmanlike approach to the job descriptions, skills, and training needed to carry out shuttle missions, with special attention to the mission specialists' portfolio. Trained to monitor, operate, and repair the spacecraft's many systems; operate the remote manipulator arm to move satellites out of (or into) the payload bay; carry out spacewalking extravehicular activities; and engage in research across a spectrum of scientific disciplines, mission specialists would constitute a very capable workforce in orbit as flight engineers and expert researchers. One reporter predicted that "it is the mission specialists who will make the reusable space shuttle . . . earn its keep."[27]

General and special interest media, especially women's magazines, covered the new astronauts more glowingly. Several of the first female astronauts became their darlings for cover stories and feature articles that examined their femininity and domesticity in addition to their professional accomplishments. Writers queried women astronauts about their diet, exercise, favorite recipes, fashion style, and attitudes toward dating, marriage, and family. Such articles indicated the problematic position of female astronauts as both a "new breed" of professional women and the traditional feminine ideal—attractive, gracious, and keeper of the home and family. Enchanted with them but not quite sure how to present them, women's magazines relied on familiar gender stereotypes.[28]

During training, it became evident that most mission specialists would not be able to maintain their own specific research projects or medical skills. They found that being an astronaut was a full-time occupation leaving little opportunity to carry out independent research, present papers at scholarly conferences, or continue practicing medicine. With some wistfulness, they accepted that they had now chosen being an astronaut as their career; they would no longer be specialists in a specific research discipline but would become highly skilled generalists responsible for doing other scientists' research in space. Acting as the orbital eyes, hands, and brains for scientists on earth, they would also become managers of varied research agendas in space. Some of their peers worried that it was frivolous to give up promising scientific careers (and higher income) to become astronauts, but the "new guys" felt the sacrifice was worth the opportunity to work in space. For many of them who remembered the glory days of spaceflight in the 1960s, becoming a scientist had been their strategy to qualify for their true career goal—being an astronaut.[29]

As mission specialists became integrated into the astronaut corps, NASA turned its attention to yet another novelty made possible by the shuttle: guest astronauts, called payload specialists. These non-career astronauts came from the business, political, and scientific communities and usually flew a single mission. They had no responsibilities for shuttle systems, only for a particular payload, such as a satellite or an experiment. This role served NASA's strategy of building support for routine spaceflight in several ways. NASA offered its commercial customers the opportunity to fly someone to tend to the company's payload. The first such payload specialist from industry, Charles D. Walker of McDonnell Douglas, spent several years developing a device for biomedical experiments and training NASA crews to use it in space before being assigned to operate it himself on three shuttle missions. NASA also welcomed aboard several passengers with lighter workloads but influence in government—a Saudi royal and a Mexican official to observe deployment of their nations' communications satellites, and a U.S. senator and a congressman who oversaw NASA's budget to gain greater knowledge of spaceflight.[30]

Scientific payload specialists flew on Spacelab missions when the shuttle carried a laboratory or observatory platforms loaded with scores of experiments. They were career astronauts from the European, Canadian, and Japanese space agencies and non-career astronauts chosen by the group of scientists whose experiments were selected for flight. The first payload specialists, one from the

United States and one from Europe, flew in 1983 on the ninth shuttle mission, the first flight of Spacelab. Payload specialists flew on research missions thereafter until the last laboratory mission in 2003. The payload specialist position gave other nations entrée to space for their own astronauts and furthered NASA's relationships with the scientific community, but it was inherently awkward because it overlapped with mission specialists. Initially, NASA's astronauts saw the role of payload specialists as redundant; mission specialists themselves were trained to operate all payloads. Nevertheless, mission specialists and payload specialists trained together for assigned missions and paired up effectively on shifts in flight.[31]

In all, fifty-one people flew as payload specialists, suggesting the possibility that spaceflight might open up for true passengers—journalists, writers, artists, educators, VIPs, and celebrities. The shuttle mission scenario had proved physically benign enough to raise the hope that "ordinary people" who were not professional astronauts might soon be able to go into space. NASA took a few steps in that direction until tragedy intervened. Designated a "spaceflight participant," teacher Christa McAuliffe was selected for flight by competing in a presidential initiative to use the shuttle as a classroom to inspire children. NASA launched a similar plan to fly a journalist in space but put it on hold and never reinstated it after the 1986 death of the *Challenger* crew and teacher. That horrific accident and the fatal end of the 2003 *Columbia* mission dispelled any hope that spaceflight was routine enough and safe enough for amateur passengers. The last U.S. payload specialist was also the first American in orbit, John Glenn; NASA extended him the courtesy of a hero's return to space on the shuttle in 1998 to participate in various experiments. The last payload specialist, and also the first Israeli in space, was fighter pilot Ilan Ramon, who perished on *Columbia* in 2003.[32]

While payload specialists sometimes caught the public's eye, mission specialists were the most noticeable face of the shuttle era. As they were primarily responsible for the activities that differentiated one mission from another, reporters naturally paid attention to them. Much of the in-flight imagery recorded them at work in the orbital lab or outside the vehicle on dramatic spacewalks, and these images graced mission coverage in print and broadcasts. The media played an active role in reinventing the right stuff by portraying astronauts as NASA did: as space workers who personified competence and diversity in the shuttle era, a professional corps that could help the shuttle reach its potential as a versatile vehicle for routine spaceflight.

Both NASA and the media shied away from treating the shuttle astronauts as celebrities, avoiding the hero worship of the first generation of astronauts.

Celebrity did not match the theme of routine spaceflight. It has been said that NASA took the most extraordinary feat—going into space—and made it ordinary by calling it routine. The shuttle-era astronauts, more than 250 in all, came to be seen as regular people—like one's classmates, neighbors, or coworkers—no matter how accomplished they were. The media generally treated them equally and rarely singled out anyone for special notice. Most were relatively unknown or anonymous outside the space program.[33]

However, shuttle astronauts still became icons, albeit of a different character than their predecessors. During the 1980s, the term "role model" entered common parlance from the field of sociology and became popular in the push for equal opportunity. A role model was someone to be emulated, mainly someone holding a professional role to which one aspired. The media readily adopted this term, characterizing astronauts as role models to demonstrate the importance of education and hard work as paths to success. Many contemporary interviews and profiles of shuttle astronauts explicitly called them role models, and astronauts' public appearances, especially in schools, embodied that purpose.[34]

NASA's announcements of the "firsts" of the shuttle era—the first American woman in space, first African American man and woman in space, first Hispanic American man and woman, and others—were temperate. As an example of the agency's acknowledgment of these milestones, the announcement of the first woman and first African American selected to fly appeared without fanfare in a press release titled simply "Three Shuttle Crews Announced." NASA left it to the media to interpret the social significance of such news. This understated approach to publicity avoided fostering a cult of individual hero worship and reinforced the shuttle-era astronaut ideal. In a diverse astronaut corps, such "firsts" would become common enough to be unremarkable.[35]

The first two "firsts" garnered the most intense media response and depiction of the shuttle astronauts as role models. Sally Ride, chosen to become the first American woman in space, tried to stay out of the limelight to remain focused on her work. When she realized that publicity was inevitable, she decided to try to shape the message: "I didn't come into this program to be the first woman in space. I came in to get a chance to fly in space." In her view, being a woman was not the story. She and the other female astronauts knew it would be important to frame the first woman's flight properly to avoid creating a burdensome stereotype. Considering themselves astronauts (not "female astronauts"), they stressed professionalism and equality. Leading newspapers noted that Ride's flight was both her achievement and a symbolic coming of age for women now ready for spaceflight. *Ms., Newsweek,* and *People* magazines made Sally Ride their cover story (plate 4). The media lavished so much attention on this historic flight (the seventh shuttle mission, flown in 1983) that commander Robert Crippen introduced the four male members of the crew as "the men who accompanied Sally Ride into space." While the media pressed for interviews and feature articles on Ride, NASA used the occasion of her flight to celebrate women's broader professional advancement by bringing together for the launch some of the nation's most prominent women.[36]

Guion "Guy" Bluford's flight as the first African American in space on the eighth shuttle mission, also in 1983, received somewhat less widespread notice beyond the black media. Profiled in the leading newspapers, he appeared on the cover of *Jet* and *Black Enterprise* magazines but not *TIME, Newsweek,* or *People;* already he was "another first." Like Ride, Bluford professed to be less concerned about being first than simply being assigned to a flight and doing his job as professionally as possible. As NASA selected crewmembers based on their readiness to meet the technical requirements of the missions, the result would naturally be evidence of the advancement of women and minorities. Noted for his reserve as a reluctant hero, Bluford realized his role in the still evolving history of blacks in aviation and as a role model for black youth, but he also believed that fame was transitory. "I anticipate that blacks in space will become more routine. So I've learned to accept it [attention] and recognize that it will eventually fade away," he remarked. "A year from now nobody will remember that 'Bluford guy.'" NASA marked the occasion of his historic flight by inviting the Tuskegee airmen and African American dignitaries to the STS-8 launch, coincidentally the first shuttle mission to launch and land at night.[37]

Popular newsstand and mass circulation magazines paid attention to the astronauts and missions on occasion, not routinely. Cover stories were rare over the course of three decades. *TIME* magazine treated human spaceflight in twelve cover stories during the shuttle era but featured only two astronauts on the cover—Michael Foale, who survived a nearly catastrophic fire aboard the Russian space station *Mir,* and John Glenn, for his return to space in 1998. *Newsweek* featured only the first mission crew (John Young and Robert Crippen), Sally Ride, Shannon Lucid, and John Glenn, on its shuttle-era covers. *People* chose only Sally Ride, Christa McAuliffe, Lisa Nowak, and the final *Columbia* crew for its cover stories.[38]

African American media enthusiastically promoted the black astronauts as high-achieving role models for their race. *Jet* proudly featured the three black astronaut candidates selected in 1978 on its cover and faithfully covered the fifteen African American astronauts who flew on the shuttle, profiling each for his or her initial flight and then following their subsequent career moves (plate 5). Guy Bluford's 1983 flight frequently appeared in the magazine's "This Week in

Black History" column as a recurrent reminder of breaking the race barrier in space. *Jet* reported on the black astronauts who had risen from humble origins and racial prejudice as well as those who had more advantages in family and education, lauding the accomplishments of all as inspiration for the black population. *Ebony* magazine likewise gave the black astronauts high visibility, especially those who accomplished firsts for the race. Mae Jemison, the first African American woman to be an astronaut, appeared often in both magazines over the years. She also appeared on the cover of an *Ebony* issue devoted to the "Year of the Black Woman" and in a *TIME* magazine special issue as "scientist of the year," and *People* listed her as one of the "50 Most Beautiful People in the World." *Essence* and black professional journals also followed the fortunes of the African American astronauts.[39]

While the shuttle astronauts concentrated on doing their jobs and NASA shielded them from too much media exposure, when the media did focus on them, reporters generally portrayed them as real people rather than idealized pioneers and heroes. Except in their hometown media and the aerospace press, few of the 250 shuttle-era astronauts received much in-depth attention other than the ones who were unusual in some way. Story Musgrave (first spacewalk on the shuttle), Bruce McCandless II (first untethered spacewalk), Kathryn Sullivan (first U.S. female spacewalker), Fred Gregory (first black pilot and commander), Franklin Chang-Diaz and Ellen Ochoa (first Hispanic Americans in space), Eileen Collins (first U.S. female pilot and commander), Bernard Harris (first African American spacewalker), Shannon Lucid (U.S. record for long duration in space), and several others received feature story treatment and some renown in the national press, but the media did not lionize them. Nor did NASA seek to promote media attention for individual astronauts, instead fostering the idea of teamwork and shared achievements.

The media, in keeping with NASA's theme of "going to work in space," typically characterized shuttle-era astronauts as workers, not explorers (fig. 3.2). They were as much the teamsters, handymen, and construction workers of space as they were brainy scientists, skilled engineers, and capable pilots. Editorial cartoonists often depicted them as auto mechanics, appliance repairmen, or plumbers—the blue-collar equivalents of their orbital roles—in a reinterpretation of the iconic astronaut. The news media generally presented a favorable image of the astronauts, but editorial pages, where proponents of less expensive robotic space explorers continued to doubt the value of astronauts, sometimes treated them more skeptically. The *New York Times,* in particular, used the derogatory term "lumbering" to describe astronauts on the shuttle whose mere presence disrupted delicate experiments and belied the alleged utility of humans in space.[40]

Fig. 3.2. Some of the astronauts' tasks in space did not require graduate degrees and test piloting skills. Cartoonists captured the humor in their more mundane chores. © Steve Greenberg for *The Daily News* of Los Angeles.

Except for the astronauts who perished in the two shuttle tragedies, none of the shuttle astronauts attained the stature of heroic cultural icons comparable to their Apollo-era predecessors. Sally Ride came close, but it was not until her untimely death of cancer in 2012 that she was widely hailed as a hero, albeit, like Neil Armstrong, a reserved and unassuming one. Even so, she had not received the Congressional Space Medal, which was awarded to earlier male astronaut "firsts" honored for exceptionally meritorious service. President Barack Obama awarded her the Presidential Medal of Freedom, the nation's highest civilian honor, posthumously; the citation praised her as a role model, and he called her "a quiet hero." Media attention to Sally Ride's death was heightened by the revelation that her life partner was a woman—the first public identification of a gay astronaut and a fact that Ride chose not to reveal while alive. Had that news come years earlier, it likely would have rattled NASA and the astronaut corps, but with more tolerant contemporary attitudes about sexuality, official and public reaction trended more toward surprise or acceptance than discomfort. Tributes flowed in from all quarters, and the lesbian-gay-bisexual-transgender community claimed a new hero.[41]

The only other shuttle astronaut to receive an unusual amount of media attention—and for opposite reasons—was Lisa Nowak, whose career collapsed in the wake of a bizarre stalking episode that led to her arrest and departure from NASA in 2007. For days stretching into months until the love triangle saga resolved, popular media fixated on the scandal in salacious detail. Magazines, tabloids, and late-night talk-show hosts mocked it as a tale of disgrace, but the national news media analytically raised questions about the stresses of being an astronaut and the adequacy of psychological monitoring and counseling. Apart from the personal trauma suffered by the direct participants, the episode so contradicted the image of astronauts as stable, self-disciplined, admirable role models that it became a public relations crisis for NASA. The agency had to respond to calls for a review of its criteria for evaluating astronauts' mental health, show some compassion for its troubled employee, and restore order by removing Nowak and the male astronaut who also figured in the morality play. This single incident shook the beyond-reproach image of the astronaut corps and revealed a human frailty that had never come so starkly into public view: like anyone else, astronauts might become vulnerable to an emotional crisis and make mistakes.[42]

For astronaut Bruce McCandless II, it was not his name or personal story that became famous—it was a photograph. One of the most recognizable and most often published images from the shuttle era was of a lone astronaut flying solo in space, suspended against blackness above the cloud-hazed blue earth (plate 6). That astronaut was McCandless on a foray in a jet-propelled backpack during a 1984 shuttle mission, but the response to the photo suggests that the spacesuited figure represents humanity at large, evidently at ease in the universe. This sublime image became iconic. It framed the astronaut as a satellite, unmoored from everything, acting as one with technology worn as a skin. The scene suggested the tranquility of space, with barely a hint of peril in the freedom and adventure of flight.

This is one of the defining images of the shuttle era, but it contrasts with the reinvented imaginary of the astronaut as a technical worker. Here the astronaut is not seen as a worker and is not identifiably McCandless unless one knows the date of the image. The astronaut is instead an explorer, free to roam at will in space. This image, like that of an astronaut standing on the moon, is likely to survive through the ages as a timeless statement about humanity, long after the memory of the actual person fades. Like icons in general, it needs no caption to communicate its significance. It is deceptive, of course, for no man is an island in space. His crewmates with the camera were close enough for rescue if necessary. Nevertheless, this astronaut image is a quintessential icon of spacefaring humanity.

Mission specialists accomplished most of the "firsts" in the reinvented astronaut corps, but the pilots broke barriers, too, and new faces appeared in the cockpit, or flight deck, of the shuttle. Frederick Gregory and Charles Bolden

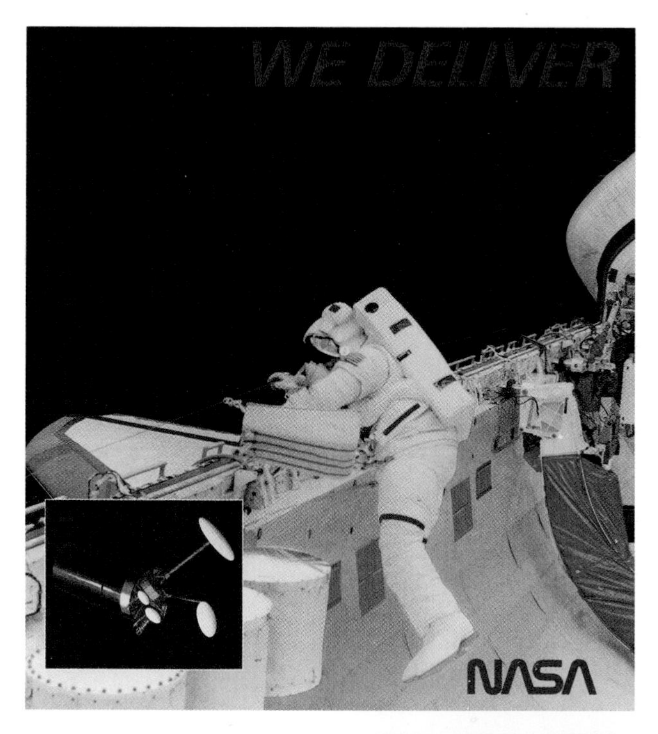

Plate 1. "We Deliver" became a motto for the shuttle astronauts, this 1983 marketing brochure, and the new era of working in space. NASA.

Plate 2. NASA issued a series of posters to celebrate the new mode of spaceflight on the shuttle: commuting to work in orbit. NASA.

Plate 3. Toy and game manufacturers introduced many versions of the space shuttle into the marketplace and promoted their educational value for "space cadet" consumers of all ages. Photo by Eric Long, Smithsonian National Air and Space Museum (NASM 2010-10691).

Plate 4. Upon selection to be America's first woman in space, Sally Ride became an instant celebrity and icon for women's ability to succeed in fields dominated by men. Reprinted by permission of *Ms.* magazine, © 1983.

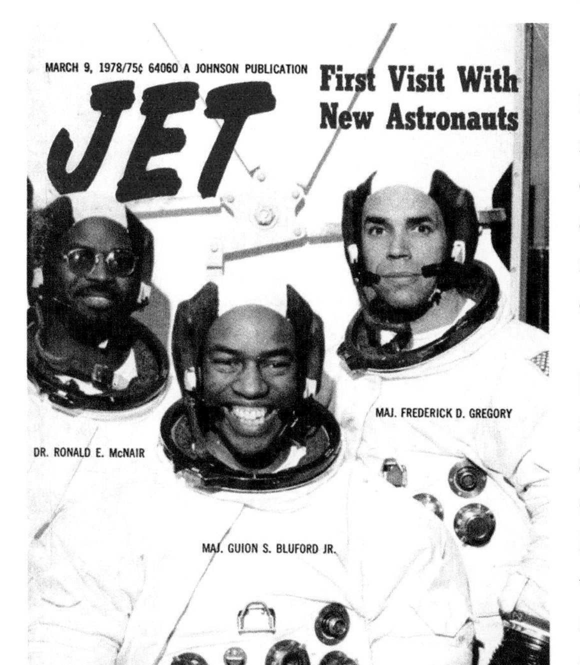

Plate 5. African American media celebrated the ascent of blacks into space and credited them as inspiring role models. Author's collection.

Plate 6. One of the most recognizable images from the space shuttle era, this photo of lone astronaut Bruce McCandless II resonates with such cultural values as freedom, individualism, and the spirit of adventure. NASA (S84-27562).

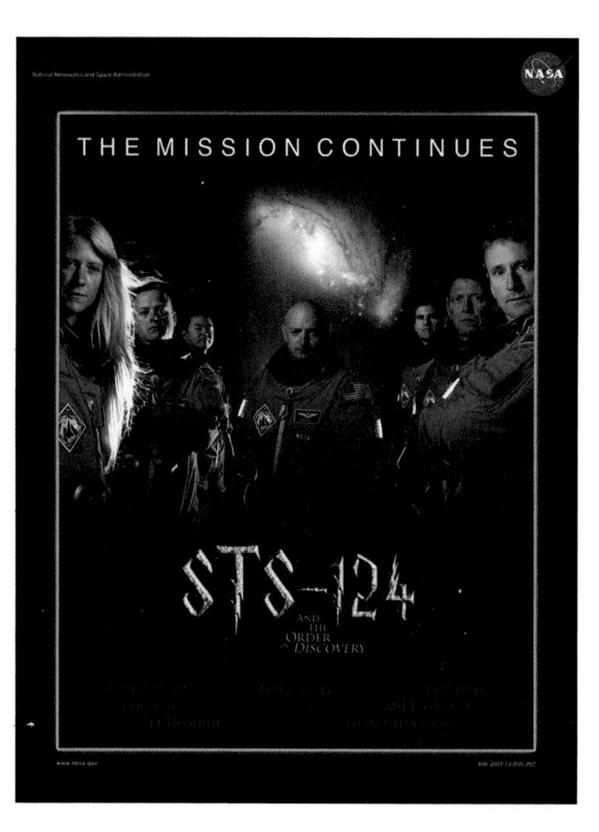

Plate 7. Some of the last space shuttle crews chose to compose their group portraits in the style of movie posters, like this Harry Potter parody, in a relaxed confluence of NASA tradition and popular culture. NASA (jsc2012e238516).

Plate 8. (*bottom right*) NASA produced a series of informational booklets about Spacelab research missions from 1983 through 1998. Cover art usually made a visual connection between the shuttle and science. NASA and Smithsonian National Air and Space Museum.

Plate 9. The microgravity symbol μg appeared prominently in the crew patch design for the last shuttle mission devoted to scientific research, STS-107 in 2003. NASA (STS107-S-001).

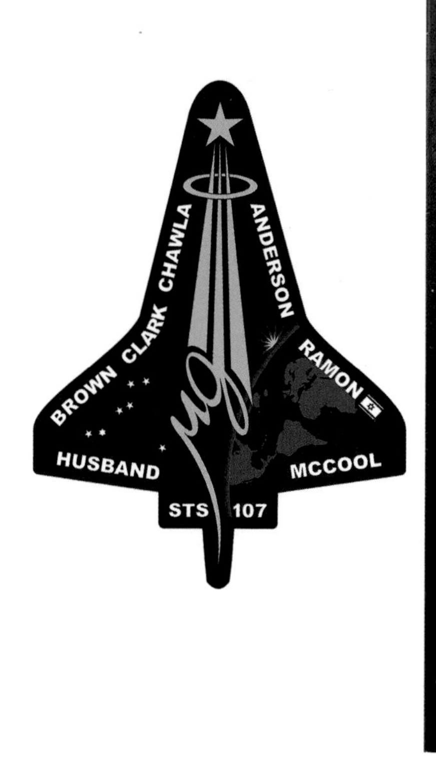

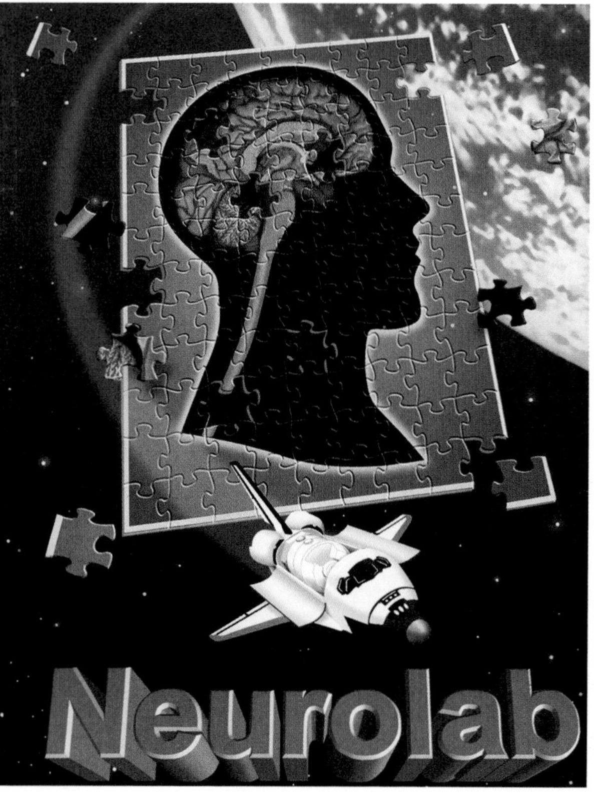

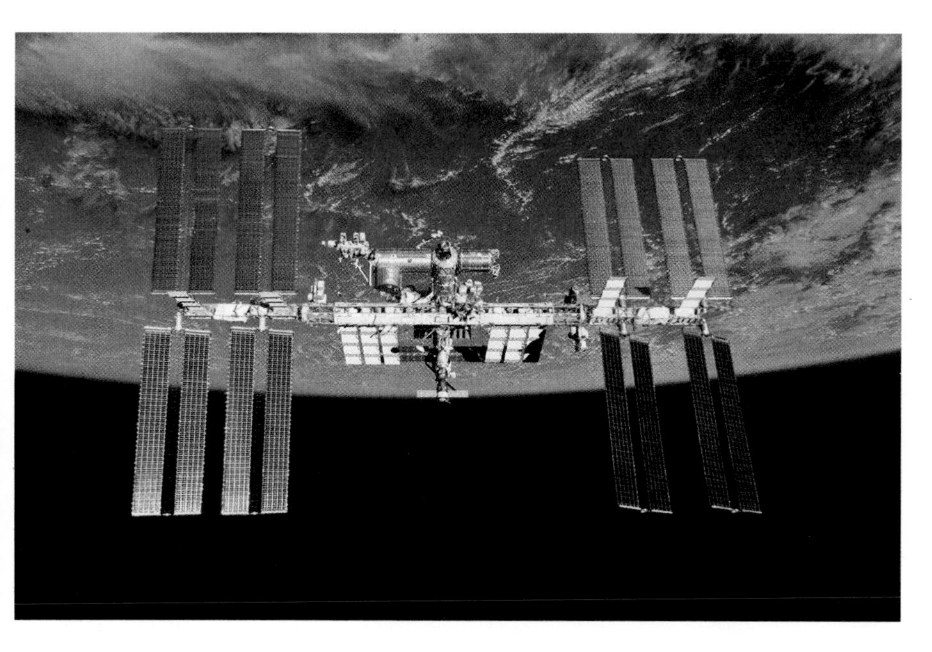

Plate 10. NASA relentlessly used "the next logical step" argument to promote the need for a space station and ultimately win approval to proceed. NASA and Smithsonian National Air and Space Museum.

Plate 11. Assembly and operation of the International Space Station—a strategic merger of scientific research and human spaceflight—became the most complex technological and cooperative venture ever attempted in space. NASA (S132-E-012208).

Plate 12. This 1986 report updated the familiar paradigm of exploring the American West with extraterrestrial landscapes and new technologies. NASA and Smithsonian National Air and Space Museum.

Plate 13. The theme of the Space Exploration Initiative—Humans to the Moon, Mars, and Beyond—is graphically depicted on the cover of this 1991 report. NASA and Smithsonian National Air and Space Museum.

Plate 14. This 2009 report doomed the Constellation program of missions to the moon and Mars as unachievable with foreseeable funding and cast doubt on future human space exploration. NASA.

Plate 15. In 2014, NASA rolled out a new motto for an iconic image that blended nostalgia for the Apollo era with hope for a human future on Mars. NASA.

Plate 16. The National Air and Space Museum presents *Discovery* as a national icon and highlights the historic firsts achieved in this orbiter. Photo by Dane Penland, Smithsonian National Air and Space Museum (NASM 2013-01353).

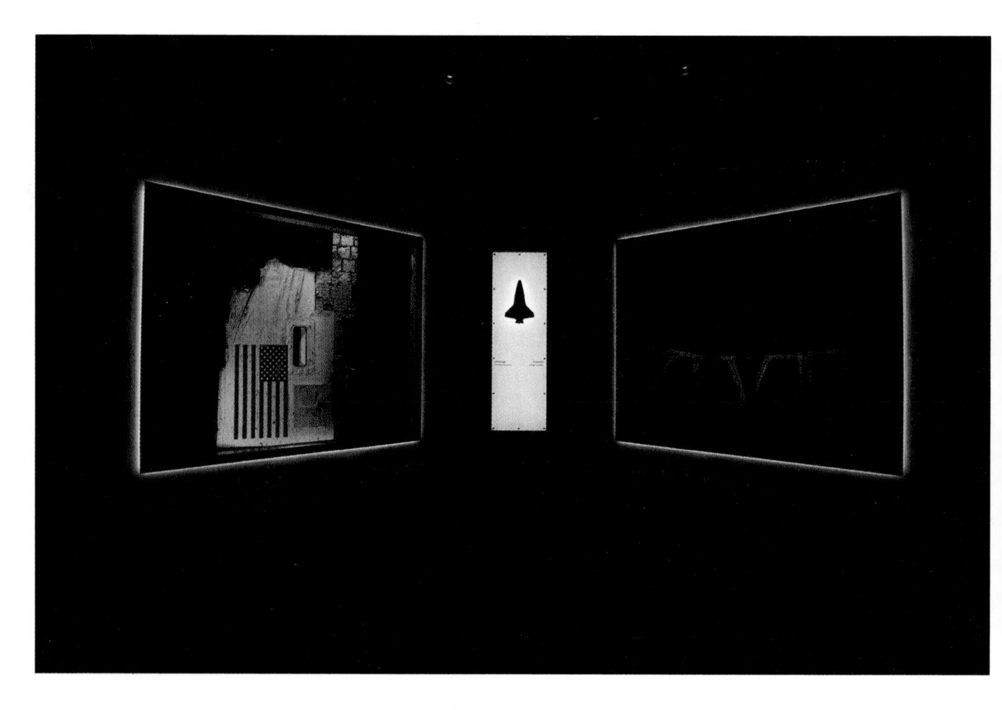

Plate 17. The *Forever Remembered* hall in the Atlantis building at the Kennedy Space Center Visitor Complex features the first public display of remnants of *Challenger* and *Columbia.* NASA/Kim Shiflett (KSC-315D-0272_0090).

became the first (and only) African Americans to pilot and command a U.S. spacecraft; they flew a total of seven missions from 1985 to 1994. The face of pilot astronauts changed even more noticeably when NASA at last chose women for that role. Because the military services did not admit women into aviation until the mid-1970s, or into test pilot school until the 1980s, or into combat aviation until the 1990s, women were not able to gain the requisite flight experience to qualify as pilot astronauts. U.S. Air Force pilot Eileen Collins, selected by NASA in 1990, came through the qualification pipeline as soon as it was possible. Her first mission as pilot in 1995 and her first as commander in 1999 made history for women and for NASA.[43]

As with the first women to enter the astronaut corps, the media strove to capture what was unique about Eileen Collins yet sometimes yielded to stereotypes. While recognizing the triumph of competently earning her way into the all-male pilot club, some reporters still commented on her marital and parental status or noted that she was shy, reserved, nice, and her crews called her "Mom." In general, though, the media hailed her ascent as evidence that space was finally an equal-opportunity workplace. Writing about Collins for *Newsweek,* Anna Quindlen observed that "the era of the firsts, the onlys, the barrier breakers, is almost gone." When the shuttle program ended in 2011, three women had managed to rise to the top as shuttle pilots. Their ascent culminated the process of reinventing the right stuff.[44]

Shuttle astronauts did not depend on NASA and the media to invent their public image and iconography; they played an active role in creating these through their own words, actions, and images. Not a mission launched without a slew of onboard cameras, so the era of living and working in space was amply documented in thousands, perhaps millions, of images taken by the astronauts. During the shuttle era, camera technology evolved from film to digital and from bulky to compact, effectively freeing crews to take as many photos with as many devices as they wished. Many shots were planned and scheduled; others were targets of opportunity and crew choices. The shuttle crew cabin was roomy enough to permit taking a variety of cameras and lenses into space, and the astronauts were well trained in their use. Their imagery, plus that recorded by NASA, media, and freelance photographers on the ground, constitutes a vast archive. These images have ethnographic value as evidence of astronaut culture. Some images simply captured moments of crew activity; others hinted of traditions and rituals; yet others reflected the diversity and versatility that characterized the astronaut corps. The content and visual rhetoric of typical images succinctly framed the idea of the shuttle-era astronaut.[45]

Some images are taken of every crew to mark certain traditions or rituals that, as is common in many organizations, serve to affirm shared values and knit a group together in shared experiences. Such behaviors and symbolism are important to esprit de corps. Imagery offers insight into both the public and private traditions that define membership in the astronaut corps. The prelaunch crew breakfast, suit-up, the walkout from crew quarters for the Astrovan ride to the launch pad, and the preflight group portraits (both official and humorous ones) are a few of the honored traditions captured in the visual record.

A favorite astronaut tradition is the in-orbit crew portrait (fig. 3.3). At some point in the mission, each crew briefly paused from work and arranged themselves for a souvenir group portrait. Usually they decided to dress alike or dress according to their shift (on round-the-clock missions, the two shifts usually wore different colors). In orbit, they usually wore NASA standard-issue pants and shorts with soft polo or rugby-style shirts in the crew's choice of colors and patterns. Sometimes they dressed for a portrait with a theme; one crew wore flowery tropical shirts, another wore dark sunglasses like the *Men in Black*

Fig. 3.3. The last *Columbia* astronaut crew was a remarkably diverse group personally and professionally. NASA (STS107-735-032).

film characters, and a space station construction crew wore hard hats. Usually the crews posed in some kaleidoscopic pattern that depicted weightlessness. Although the crews took these photos for themselves, the images were often included in NASA's selection for public release. In these unique in-orbit portraits, the astronauts took control of their own image, and the public sometimes had a glimpse of them in a more relaxed mode.

These informal portraits celebrated teamwork. Crews trained together for a year or more, learning to support and depend on one another like a family. The portraits also celebrated diversity—both the evident diversity of male and female and racial and ethnic heritages and the invisible diversity of military and civilian, scientist and engineer, and innumerable differences in backgrounds and interests. Shuttle crews became cohesive and effective teams, as the in-orbit portraits portrayed. A shuttle mission was not complete without that memento.

The official crew portraits made on the ground before the mission were variations on a formal template. Except for the different faces, these looked alike, with only slight differences in poses, props, and backgrounds. They followed traditional conventions of portraiture and were easy to "read." The crew usually arranged themselves with the commander and pilot in front or center of the rank of specialists. The United States flag usually framed the left margin of the image, and a NASA flag or a foreign crewmember's national flag framed the opposite edge. A model of the space shuttle, Hubble Space Telescope, or another mission-related payload might be held or placed in view, or a scene related to the mission—a celestial view or a launch, for example—might appear in the background. Most of the crews wore their blue or orange launch-entry suits, with the spacewalking astronauts in white extravehicular activity (EVA) suits, all sporting their mission's patch.

These formulaic portraits, though similar, represented the astronauts' effort to define themselves by including the props that symbolized their distinctive mission and sometimes subtly signaled their values, as when the commander was not positioned front or center. The official formal portraits had many uses; reproduced in press kits and as mission posters, they made their way into the media. As the most common still images of astronauts in the public domain, these portraits became symbolic of the shuttle-era astronaut corps.

A new type of crew portrait appeared late in the shuttle era and became quite popular. NASA's Space Flight Awareness office had for decades produced posters based on the official crew portraits and distributed them to the spaceflight workplaces. These posters served as graphic reminders of the employees' responsibility for crew safety through quality workmanship. Like the crew portraits, the posters tended to follow a similar pattern and look more alike than original. For the post-*Columbia* accident return-to-flight missions, an engineer with an artistic bent created something different as homage to the excitement of spaceflight and many workers' interest in space science fiction. At the request of an astronaut, he designed one with some of the visual elements of a movie poster ("Coming Soon," "The Event of the Year," and the astronauts' names stacked like film credits).[46]

This poster became a sensation within the shuttle workforce and, courtesy of the internet, quickly reached the public. Soon thousands were printed and in circulation. The parody proved so successful in catching attention and creating buzz that NASA launched a series of mission "movie posters" inspired by the promotional posters for various popular films (plate 7). Many of the shuttle missions and space station expeditions from 2005 on portrayed the crews in the style of such popular films as *Star Trek, Star Wars, Harry Potter, Indiana Jones, Armageddon, The Matrix, Pirates of the Caribbean,* and others. The genre expanded to parodies of book jackets (*The Hitchhiker's Guide to the Galaxy*) and record album covers (The Beatles' *Abbey Road*). The posters were greeted as one of the coolest things staid NASA had ever done. By depicting the brainy astronauts in the guise of popular culture, they made a lively new connection with the public.

Among the most stunning images from the shuttle era are those of astronauts working outside the spacecraft. Extravehicular activity, or spacewalking, in earth orbit does not involve walking as it did on the moon, where one-sixth gravity held men and objects on the surface. Astronauts working around the shuttle and International Space Station are in free fall and need their legs only to anchor themselves somewhere. They work primarily with their upper bodies, arms, hands, and heads. Much of the time one astronaut is positioned on the end of the fifty-foot-long robotic manipulator arm, being moved around by another astronaut operating the arm from inside the shuttle or station. EVA astronauts work in pairs, spending up to eight hours outside doing a variety of tasks, always tethered to the spacecraft so they do not drift away.

Images from two sets of missions framed especially well the astronaut as skilled worker. Five times astronauts returned to the Hubble Space Telescope, more than three hundred miles above earth, to repair and refurbish the bus-size observatory to extend its operational life. The first of those visits saved the telescope and the agency from disgrace by correcting the optical error discovered during in-orbit testing. NASA's reputation rode with that crew. Servicing crews' tasks ranged from exchanging instruments the size of telephone booths and baby grand pianos to doing surgery to extract circuit boards from electronics boxes, using some 150 tools for the job. By the time they completed their last task, the twenty-year-old telescope was essentially new again. Images from the Hubble servicing missions showed the astronauts at work in odd settings, squeezed inside the observatory's bays, clinging like insects to its sleek tube, lifting the large solar arrays and boxy instruments with apparent ease. The capability to service satellites in space had been a selling point for the shuttle, and the astronauts proved their value repeatedly.[47]

Another set of images framed the astronauts as orbital construction workers. They assembled the International Space Station in orbit, module by module, truss by truss, in a series of almost forty shuttle missions. Only the images made intelligible how large the station is and how it dwarfed the astronauts working on its appendages. Some images showed an astronaut far out on the end of a robotic arm with nothing in view but the earth below; others were close-ups of astronauts' gloved hands deep in the bowels of a power system full of cables and connectors. These photos gave a sense of both the force and the finesse—and the boldness—required to do work in space. The capability to build a space station had been a selling point for the shuttle, and the astronauts did it. Both the Hubble servicing and space station imagery became emblematic of the dramatic expansion of human capabilities in space and built confidence that people would be able to set up a base on the moon or Mars if such a goal were adopted.

The visual rhetoric of EVA images was less formulaic than that of the crew portraits, but some elements appeared often enough to become recognizable symbols of the astronaut as worker. Tools figured prominently in many EVA photos, either in the hands of an astronaut or tethered to a wrist or work station for ready use. The tools seen in pictures from the Apollo-era EVAs were primarily for geological tasks; those seen in the shuttle era were mostly meant for installation and repair tasks—wrenches, pliers, caulk dispensers, and cordless power tools like those in common use on earth, plus some unfamiliar custom tools. During the shuttle era, EVA tasks grew ever more complex and challenging, and the astronauts expanded the range of work successfully performed in space. Women astronauts proved their mettle as spacewalkers, handling small tools and massive equipment and using the robotic arm as adeptly as the men. Two by two, and once with a third crewmate, EVA astronauts performed their well-choreographed tasks. Regardless of jokes and cartoons about space-age mechanics, the astronauts proved to be highly capable servicing experts.

The EVA suit also figured prominently as the work uniform of orbital servicing and construction workers. Looking somewhat like a protective hazmat suit, the bulky white Extravehicular Mobility Unit (pressure suit) worn by shuttle astronauts signaled the extreme environment of their worksite. In reality a technically complex personal spacecraft, the suit provided life support with enough range of motion and comfort to permit eight-hour workdays outside. Unlike the custom-made suits of yore, the shuttle EVA suit consisted of parts—arm and leg segments, shoulders, waist, and chest units in different sizes—that could be assembled to fit almost any astronaut except the smallest women. Like the orbiters, the suits were interchangeable and reusable, and in a sense, so were the astronauts. In an era of routine spaceflight, custom-fitted suits for every crewmember would have been a luxurious expense, so they shared ready-to-wear garments. In combination, tools and spacesuits epitomized the astronaut as one who does useful work in space. EVA images offered a compelling visual rhetoric for the reinvented astronaut of the shuttle era.

Even more than still images, the IMAX motion pictures filmed partly in space by the astronauts made spaceflight vivid. With special training, advance development of storyboards, and meticulously planned shot schedules, selected shuttle crews used two 70-mm large-format IMAX cameras (one inside and another mounted in the payload bay), and later a 3D IMAX camera, to record their activities and views of earth as faithfully as possible. *The Dream Is Alive,* the first of seven IMAX documentary feature films made in cooperation with NASA, was released in 1985. It introduced the space shuttle and its astronauts to millions around the world. *Blue Planet,* a dazzling tour of earth as the astronauts see it, was released in 1990, followed by others featuring astronauts on *Mir,* the International Space Station, and Hubble servicing missions. These films of the shuttle era effectively brought its achievements and astronauts down to earth and heightened awareness of human spaceflight.[48]

Projected on a screen several stories tall, the films in this series offered viewers an immersive experience, almost as if they were in space themselves. Audiences enjoyed virtual face-to-face contact with the astronauts in training and in flight. They almost "felt" weightlessness as the astronauts glided across the crew cabin or moved around on spacewalks. Through the vicarious IMAX experience, anyone could begin to understand what it was really like to be an astronaut in the shuttle era. Produced to be shown in museums and other educational venues, then later marketed in formats for viewing in homes and schools, these films were lauded, even by astronauts, as a highly realistic experience compared to actual spaceflight. They succeeded in telling good stories that informed, awed, and excited people about spaceflight.[49]

The vast body of still and motion imagery, whether documentary in nature or posed for intentional effect, played a role in reinventing the astronaut ideal. Images transformed the astronaut imaginary from heroic pilot to a more populist version reflecting the new demographic. This visual culture represented and helped popularize the multiracial, multiethnic population of women and men, scientists and engineers, who embodied NASA's theme for the shuttle age—"Going to Work in Space."

But going to work in space was not always about smiling astronauts and pleasant traditions. Twice during the shuttle era Americans and the world were shocked by the tragic loss of a shuttle and its crew, once while leaving earth and once while returning. The attitude of NASA and the public toward the astronauts took a marked turn in the wake of the *Challenger* and *Columbia* tragedies in 1986 and 2003. When seven crewmembers died on each of those missions, NASA and the nation temporarily abandoned the trope of astronauts as workers and valorized them as space explorers and heroes. In the rhetoric and iconography of mourning, these astronauts became extraordinary individuals, elevated and reframed in a tragic narrative impressed on public memory. Their deaths prompted fresh public discussion about the risks and value of human spaceflight and, at least briefly, put astronauts in the public's consciousness. The fact that both crews were among the most visibly diverse of all shuttle missions and thus especially representative of all Americans may partly account for the widespread mourning. The death of astronauts became occasions to reconsider the place of space exploration and the astronauts' status as heroic icons in American culture.[50]

The seven-person crew of STS 51-L, *Challenger*'s last mission in 1986, included two women and five men, one an African American and one an Asian American. Five were astronauts, one was a corporate payload specialist, and one was a teacher chosen as a spaceflight participant; three were on their first flights. Their mission to deliver a satellite lasted only seventy-three seconds until ending in an explosive breakup during ascent; they died without reaching space or realizing the potential for which they had prepared. The public responded especially to one crewmember's poignant story. Christa McAuliffe was not a professional astronaut, yet she was a person with whom everyone could identify, a schoolteacher, wife, and mother whose passion for teaching had earned her a competitive seat on the mission. In an already diverse crew, she was the one who captured widespread attention.[51]

The seven-person crew of STS-107, *Columbia*'s last mission in 2003, also included two women and five men, one an African American. All but one, a payload specialist from Israel, were professional astronauts; four were on their first flights. Their very successful scientific research mission lasted almost sixteen days and ended just sixteen minutes short of landing when the vehicle, damaged during launch unbeknown to them, disintegrated in the atmosphere. *People* published their joyous walkout photo on its cover under the banner "Brave Hearts," and their in-flight portrait, found in film recovered from the debris, froze them in time happily floating together, all smiling. The public responded especially to two of the crewmembers' stories. Ilan Ramon, a devout Jew, a decorated fighter pilot, and national hero, was the first person from Israel to fly in space. Kalpana Chawla, a native of India, had improbably fulfilled her childhood dream of going into space by moving to the United States for her engineering education and becoming a citizen. Many media reports accented these two distinctive crewmates.[52]

There certainly were other crews with a similar mix of gender, racial, or ethnic identities, and national origins, and there also were some much more homogeneous crews. That the two mission tragedies took notably diverse crews highlighted the reality that spaceflight remains a high-risk endeavor that may claim anyone. The death of McAuliffe especially struck a chord, because everyone perceived her as an "ordinary person," a representative and symbol for the public, and her mission was meant to show that spaceflight was safe enough for everyone. Even more than the new breed of astronauts, she symbolized the new era in space.[53]

Other astronauts had perished before in jet crashes and car accidents and in the horrific Apollo spacecraft fire, but never in such public circumstances. The launch crowd and live television audience had directly witnessed *Challenger*'s disintegration seconds into its ascent, and people along *Columbia*'s descent path over two states had seen and heard its debris fall from the sky. Vast audiences became virtual eyewitnesses to the deaths of the shuttle astronauts as the events were replayed incessantly on television. Viewers gained a sense of familiarity with astronauts about whom they had known little or nothing hours before. Intense media coverage captured on-the-scene expressions of shock, grief, and public reactions from many locations, creating a huge virtual community of mourners that bonded the nation in bereavement. "America wept," *TIME* reported.[54]

Two images, first seen on television, instantly become iconic symbols of the astronauts' deaths and distillations of the nation's grief. *TIME* and *Newsweek* ran them as cover photos of *Challenger*'s explosive breakup and *Columbia*'s fragmentary return. A viewer who had not seen the barrage of televised news might not immediately recognize the significance of these images—one appearing to be a strange fireworks display and the other a cluster of meteors streaking across the sky—but to the informed they were almost unbearable symbols of catastrophic failure and loss. The two magazines adopted different visual rhetoric on their covers. *TIME* presented each image as a portrait trimmed in black, with only a brief memorial caption: "Space Shuttle Challenger, January 28, 1986," and "'The Columbia Is Lost,' February 1, 2003." These covers were compositions in red, white, blue, and black with flashes of light. *Newsweek* framed the images less artfully with verbiage, pairing the *Challenger* photo with titles of related articles—"What Went Wrong?" "The Future of Spaceflight," "The Shuttle's Heroes"—and adding the crew's portrait to the meteor image. Years later, these images need no caption for those who mourned the astronauts, and they continue to trigger a shared cultural memory as symbols of those awful events.[55]

Public mourning in response to national tragedy and disaster is an important aspect of healing and memory. Until recently the primary occasion for such mourning was the death of a head of state marked by ceremonies in which the public could participate by actual attendance or experience in absentia via the media. Increasingly the public has responded to other prominent deaths and is not content to file passively by a memorial. Many want to engage more actively in mourning and commemorating significant losses and in shaping the public memory of those events. Response to the space shuttle tragedies suggests how invested the public remains in the astronauts' role as heroes, despite their reinvented role and image in the shuttle age.[56]

The public rituals of mourning evidenced a spontaneous reframing of the role of astronauts as heroes to make sense of the tragedies. In response to the news, people began to make somber pilgrimages to engage in acts of mourning. Thousands went to the "homes" of the space program—NASA centers, the National Air and Space Museum, and other space museums around the country. Many came with a sense of purpose to perform an act of tribute or leave a token of remembrance, respect, or condolence that briefly hallowed these civic places as spontaneous shrines. Others made impulsive offerings when they observed the objects and messages collecting there, making a decision to join in the ritual. Thus began the public mourning for the astronauts and the affirmation of their place in public memory.[57]

The items left at these shrines—flags, flowers, teddy bears, NASA memorabilia, religious tokens, notes and cards, and more—became the material culture of grief. They suggested how people were interpreting the meaning of these tragedies and attempting to reconcile the astronauts' deaths with the idea of routine spaceflight. Studies of objects left at memorial sites indicate that such mementos represent beliefs and values; otherwise ordinary tangible objects became expressions of sorrow, patriotism, faith, and comfort. For the shuttle tragedies, these mementos and informal acts of mourning revived the notion of heroism at a local and personal level.[58]

The cycle of official mourning ran its course, beginning with statements by the president, NASA officials, and other public figures. They established a heroic rhetoric of praise for the astronauts as courageous and inspiring servants of a national purpose. For days, flags flew at half-staff and civic life was somber. The formal mourning culminated in televised memorial services, eulogies delivered by the president, and ceremonial honors accorded to national heroes. Official mourning ended with calls for a rededication to space exploration as the most appropriate way to honor the astronauts' sacrifice and carry on their legacy.

After any national disaster, the president's first duty is to offer comfort and reassurance, and Presidents Ronald Reagan and George W. Bush rose to the occasion for the shuttle tragedies. Presidential rhetoric in the midst of such trauma is crucial to help the public understand what has happened and how the nation will respond with dignity. Speaking as mourner-in-chief, "interpreter-in-chief," and symbolic head of the national family, the president must heal and unite the nation, assuage its anguish, and offer hope for the future by framing the event to establish its meaning. President Reagan did exactly that in a televised address from the Oval Office barely six hours after the *Challenger* loss.[59]

Reagan's speech formally reinvented the astronauts, elevating them as heroes and admitting that the nation had grown complacent about the risks of spaceflight. He recognized them as actors in the defining national myth of exploration and pioneering, using metaphors that already marked his presidential rhetoric. He spoke of courage, faith, hope, freedom—many of the eternal touchstones of civic life and his presidency. He embraced the families, the schoolchildren who had watched the launch gone terribly awry, and NASA's workers with the nation's sympathy. He offered meaning in the present tragedy by setting it in a larger historical context, as an occasional and awful consequence of exploration, and he made a promise that the journey would continue. Without the words death or disaster, the speech moved from shock to mourning, then to praise and acceptance, and finally optimism and hope. In mourning the astronauts, Reagan reminded the nation of its history and framed the public memory of this day's event. Not since President Kennedy had the nation's leader affirmed space exploration, the frontier, and the future together so eloquently. And, in a matter of minutes, the speech displaced the horrific scene of the *Challenger* mission's explosive end with the memory of the crew's joyful departure. The lead speechwriter wrote that neither she nor the president realized how powerful it was until the next day, when an avalanche of favorable mail, calls, and telegrams arrived. Three days later, President Reagan delivered the eulogy for "our astronauts" at the outdoor memorial service at the Johnson Space Center.[60]

President George W. Bush made a similar, though briefer, address on the day of *Columbia*'s loss. He also sought to comfort the families and the nation, and he also vowed that the journey in space would continue. The president noted that it is easy to forget the risk when spaceflight seems routine. Without using the word heroes or heroic, he recognized the astronauts' courage and daring. Bush offered comfort from scripture, and his message ended with the rhetoric of faith. Like Reagan, he also delivered the eulogy for NASA's memorial service in Texas and honored the astronauts as heroes. On both occasions his words satisfied the same presidential role to memorialize the astronauts, comfort the grieving, find meaning in untimely death, and reassure the nation about the future. Whereas Reagan expressed faith in space exploration, Bush turned to religious faith for meaning. Less than a year after remarking on exploration and discovery as destiny in his eulogy for the astronauts, however, this president decided to end the shuttle program and propose a new journey in space.[61]

As the presidents' words raised astronauts to heroic status, editorial cartoonists used imagery to do likewise. They rendered grief in powerful images that spoke to the national mood. According to cartoonist Daryl Cagle, when emotions are strong, it is a challenge to avoid clichés. Cartoonists drew upon a small set of symbols to convey their responses to the deaths of astronauts (figs. 3.4 and 3.5). Their drawings often reflected the same values as the mementos left at spontaneous shrines. Cartoonists' visual rhetoric included Uncle Sam or the American eagle in mourning—shedding a tear, looking morose, head bowed in grief, saluting, or waving goodbye. A personal variant of this patriotic imagery showed a child or family weeping. A more spiritual rhetoric depicted the astronauts as stars in the sky, scattered or forming a constellation, or as stars fallen from the U.S. flag. Occasionally two symbols were combined with Uncle Sam and seven stars in the same frame or the flag at half-staff against a starry sky. Yet another approach integrated some element of the tragedy, such as the meteor pattern of *Columbia*'s descending debris, into the flag. For the first days after each tragedy, cartoonists held their barbs and offered the touchstones of patriotism and a generic spirituality as condolence for the nation's grief. Later, as the accident investigations progressed, the barbs returned in cartoons meant to provoke, not to comfort.[62]

The media bore witness to these first days of mourning and memorializing, but also pressed ahead for answers and explanations. Immediate news coverage, of course, focused on the visible events of the two mishaps and tributes to the fallen astronauts, but very quickly the agenda shifted and investigative stories began to dominate. NASA, a presidential commission, and the media launched

Fig. 3.4. Cartoonists drew upon a variety of patriotic symbols to convey the nation's grief at the loss of the *Challenger* crew in 1986. © Doug Marlette Estate.

Fig. 3.5. This response to the demise of *Columbia* and crew transformed the meteoric streaks of the disintegrated shuttle's debris into a patriotic image suggesting heroism. © 2003 King Features Syndicate, Inc., World Rights Reserved.

inquiries into the causes of each accident. Journalists probed into deliberations, records, and anything else that might explain what had happened and why, and whether the problem could be fixed. They raised troubling questions about lax regard for crew safety, lingering hardware flaws, the astronauts' real awareness of risks associated with such problems, management and decision-making, and beyond that the very future of human spaceflight. They extrapolated from tragedy to the strategy and priorities of the space program, questioning whether it was on the right path. "Should U.S. Continue to Send People into Space?" read a headline just two days after the *Challenger* disaster; the question arose again in 2003.[63]

After respectfully acknowledging heroic lives lost on the shuttle, leading editorial pages unleashed frustrations with NASA's shuttle program and its style of human spaceflight. When routine spaceflight turned catastrophic, it invited negative commentary and a focus on risk. The argument that unmanned exploration was far more productive and far less risky and expensive than human spaceflight moved to the fore, and the opinion that NASA needed a new goal worth the risk of human life gained traction. As evidence of faulty decision-making mounted, so did rhetorically strident calls to abandon "the delusion" of routine spaceflight and "the folly" of risking lives on such routine work as carrying satellites into orbit. In the months after the *Challenger* loss, editorialists called for rethinking space policy and the priority of human spaceflight. The *New York Times* repeatedly suggested a joint U.S.-Soviet mission to Mars as a goal worthy of sending humans into space, and Carl Sagan offered there a multi-pronged Mars exploration project to revitalize the space program still adrift a year after the tragedy. Building a space station trumped a mission to Mars in the years between the two tragedies, but advocates revived the Mars argument with more success after the loss of *Columbia.* These demands for a new goal for human spaceflight in effect called for reinventing the astronauts again, this time for a future as explorers beyond earth orbit. Death and destruction prompted a return to the original astronaut imaginary, a vision of humans properly in space as explorers.[64]

John Noble Wilford, the *New York Times* science chief who had introduced the space shuttle to readers almost ten years earlier and reported its missions since then, took a broad view of the meaning of the *Challenger* tragedy. He barely mentioned the crews but turned his attention to the bigger picture of human spaceflight and U.S. prominence in space. His last column of 1985 noted that the coming busy year, with fifteen planned shuttle missions, was crucial to NASA's plans for the future; it might finally defuse some of the criticism that the shuttle was unreliable and not yet routine or economical. The "gathering momentum" of shuttle flights would be important in moving toward approval of a space station and plans beyond. A month later, Wilford was meditating on the reality of risk and a complacent faith in technology; death forced the realization that space travel was not yet routine. As he reported news of the investigations and doubts about NASA's "fabled competence," he continued to ponder what the tragedy would mean for the future. At a bleak time, he chose to be confident that astronauts would fly again, asserting that NASA would not let the shuttle fail, even with a reappraisal of its priorities, nor would the United States retreat from leadership in space. He greeted the *Columbia* tragedy with similar questions about technology, risk, and the value and future of human spaceflight but with diminished regard for the shuttle.[65]

Through the experience of the shuttle tragedies, the lexicon of human spaceflight shifted. Dark words now came to mind: disaster, catastrophe, horror, tragedy, explosion, fireball, disintegration, debris. This vocabulary of shock, wreckage, and death served as a reminder of risk and danger, the largely unspoken realities of launching people on rockets and streaking home from space. The two shuttle accidents almost killed the idea of routine spaceflight for going to work in space, and certainly left it damaged. Loyalists ("shuttle-huggers") had to admit that the shuttle-era ideology was too innocent and complacent; critics and cynics charged worse, that it had been disingenuous from the start. The shuttle's successes—*Challenger* launched on the 25th mission and *Columbia* on the 113th—lulled the public into taking it for granted. With the shuttle grounded after each mishap and investigators drilling to find fault, U.S. human spaceflight might well have ended. Astronauts served in many roles during the accident investigations, essentially working to save their profession.[66]

Upon *Columbia*'s loss, the term "experimental" replaced "routine," as in "the shuttle has always been an experimental vehicle." Yet, when flights resumed with new safeguards, to the public they looked much the same. Astronaut crews shuttled back and forth to assemble and supply the International Space Station twenty-two more times and serviced the Hubble Space Telescope one last time. Crews resumed working in space as engineers and scientists, seeing themselves not as heroes but as people with great jobs. As Presidents Reagan and Bush had promised in the eulogies, nothing ended; more shuttles and astronauts flew, and human spaceflight survived. A cumulative 110 missions followed the tragedies and accomplished the pledge to honor the memory of shuttle heroes by continuing the work to which they had dedicated their lives. Human spaceflight became a living memorial that their deaths "were not in vain."

As the cycles of official and spontaneous mourning ran their course, the process of permanent commemoration began. As cultural products, monuments and memorials are meant to preserve and communicate the narrative formed in public memory. NASA, the crews' families, and many communities and organizations decided how to create a worthy public legacy to ensure that the shuttle heroes would not be forgotten. Monuments of various types appeared in parks and civic spaces around the United States, and in India, Israel, and elsewhere. Schools and streets, buildings and highways, mountain peaks, craters on the moon and Venus, Mars landing sites, and asteroids were named for the deceased astronauts, singly or as a crew. Scholarships, awards, academic positions, and charitable foundations also bore their names. Such memorials satisfied a need for enduring public memory as a hedge against the forgetful march of time.

Shortly after each accident, the government enacted legislation to establish a national memorial to the crew in Arlington National Cemetery, where so many of the nation's heroes are laid to rest. Each was dedicated close to the first anniversary of the crew's death, and the two sculpted bronze plaques on marble markers stand together there. A third national memorial, dedicated in 1991 by NASA and the Astronauts Memorial Foundation, rose at the Kennedy Space Center Visitor Complex. The names of the *Challenger,* and later the *Columbia,* crew, and those of other astronauts who died on duty, were carved through a huge Space Mirror Memorial—a polished black granite wall that reflects the sky as sunlight sparkles through the names. The Johnson Space Center added seven trees, and then seven more, to the memorial grove where NASA remembers its family of astronauts. NASA added a late January Day of Remembrance to the agency calendar for observance at each of its sites.[67]

The families of the *Challenger* crew aimed for an active living memorial to continue the educational thrust of their loved ones' mission. They established a foundation to fund an educational center that ultimately became a series of Challenger Centers for Space Science Education, where students and teachers participate in mission simulations and other immersive learning activities. By 2015, forty-five Challenger Centers in the United States and three other countries had introduced millions of learners to the challenges of spaceflight. Programs link educators and young people with astronaut role models to pass the spaceflight legacy on to another generation.[68]

In 2004, President George W. Bush bestowed the Congressional Space Medal of Honor posthumously on the members of the last *Challenger* and *Columbia* crews, a further emblem of their status as national heroes. As of 2016, only three other shuttle-era astronauts had been accorded that honor: John Young and Robert Crippen for the first shuttle mission, and Shannon Lucid for her record-setting stay on Russia's space station *Mir.* Unlike the NASA Space Flight Medal presented democratically to all astronauts upon completion of their missions, this medal is selectively awarded upon NASA's recommendation. It is NASA's prerogative to single out from an egalitarian astronaut corps a few that it deems deserving of special recognition.

Living astronauts also may be accorded status as heroes. Each year, a few no longer on active duty are selected for induction into the United States Astronaut Hall of Fame in Florida. Ostensibly meant to honor astronauts with a display of their accomplishments and memorabilia in a Hall of Heroes, this organization also collaborates with the Astronaut Scholarship Foundation to support educational opportunities. The existence of the Astronaut Hall of Fame and its ceremonial inductions suggest the lingering desire, in the public and in this organization's membership, to have astronaut heroes to honor and remember.[69]

The idea and iconography of heroic status did not die in the shuttle era, despite the reinvention of astronauts in the image of workers. The tragic, public deaths of astronauts triggered their elevation as heroes, but that did not happen for those who died more privately off duty, of disease or natural causes, or for those who left the agency for retirement or other pursuits. This disparity suggests a fluid standard for heroism in astronauts. Being an astronaut, while admirable, is not tantamount to heroic status. Normally, for human spaceflight, the rhetoric and symbols of heroism are latent until a crisis elicits them. That is when complacency is shattered, risk is recognized, and the nation reaffirms its desire for heroes. An editorialist once commented that "dead astronauts are always heroes" for two reasons: because of their bravery and also because heroism elevates spaceflight out of the mundane. As the nation's surrogates in space, astronauts become what the public needs them to be—role models, explorers, and sometimes heroes.

When the shuttle era officially began with the first launch in 1981, the astronaut corps numbered about eighty people: twenty-seven holdovers from the Apollo era and fifty-four new recruits (Groups 8 and 9). By 2012, when the shuttle program ended, NASA had repeated astronaut recruitment twelve more times, at one- to five-year intervals. In all, 257 more pilots, scientists, physicians, and engineers entered the astronaut corps from across the nation. They were white, black, brown, Hispanic, Asian, Native American, Indian, male, and female. A few were naturalized U.S. citizens. Shuttle astronauts came from diverse backgrounds, universities, academic disciplines, and all branches of military service. All were well educated and highly motivated to succeed.[70]

Early concerns about female physiology and temperament proved baseless; women showed they were healthy and productive members of mission teams.

Shuttle astronauts proved that gender made no difference among professionals; women succeeded in training and executed their roles in space as capably as men. About the only noticeable difference in space was a requirement to confine billowing halos of hair in clips or ponytails as a matter of safety and courtesy, a small emblem of changes that occur when women enter a new workplace. In all, forty-eight different women flew on shuttle missions; thirty-two times, the crew included more than one woman, and three times, three women served on a crew together. A couple of times the media speculated about the possibility of an all-woman crew, but female astronauts tended to dismiss it as a stunt unless mission objectives required women only for gender-specific research. Around 2000, the idea had some support at NASA as a way to focus on women's health issues, but no such recommendation came from an advisory panel on biomedical research.[71]

From the mid-century post–World War II baby boom and then from the Generation X cohort, many had grown up fascinated with spaceflight and determined to become astronauts, some applying several times before being selected; yet others dared not even imagine such a career until it became possible for them to apply. Sally Ride said that it dawned on her only when she read the recruiting ad and realized, "I'm one of those people." As they were assigned to mission crews or made public appearances around the world, shuttle astronauts personified the idea of diversity. Still among the best and brightest products of America's educational systems, they also reflected the potential for anyone who was qualified to rise into this demanding and admired profession.[72]

As "ordinary" people with "extraordinary" jobs, astronauts in the shuttle era were well rounded, with interests and talents beyond their technical specialties. As their NASA biographical summaries indicate, many excelled as competitive athletes or talented musicians and performers. Others earned multiple advanced degrees. Their jobs required public appearances around the world, and these were opportunities to present the values of the astronaut corps in their own words. They answered questions about spaceflight and talked about their life journey toward space, the importance of doing well in school and meeting goals, the value of teamwork, and other virtues. Upon leaving NASA, many of these astronauts continued to be visible as motivational speakers, educators, consultants, business executives, and authors.[73]

As U.S. public education put increasing emphasis on the STEM disciplines—science, technology, engineering, math—astronauts became the role models to emulate. Their dedication to excellence was as important to the public image and inspirational value of astronauts as their technical expertise was to their jobs. Most appeared to be personally modest, yet proud of their role in making spaceflight a routine and productive aspect of American life. They talked about the importance of space exploration but did not particularly see themselves as explorers. They were doing their jobs in space near earth to prepare for a possible future of human journeys afar.

For all their accomplishments in 135 missions, shuttle-era astronauts never left earth orbit. Cumulatively, they traveled billions of miles in a circuit, trapped on a speed-limited beltway around earth without an exit to the moon or Mars. Their space truck was not built for space exploration in the truest sense; it was built for work. Thus they were space explorers only in a constrained practical sense, as in exploring how to live and work in space. However, in one domain they were indeed pioneers—exploring the frontiers of knowledge. The shuttle was their orbital laboratory and observatory. As researchers, the subject of the next chapter, astronaut crews explored some of the mysteries of science by examining, in microgravity, phenomena that cannot be studied well on earth.

Chapter 4 Science: Doing Research in Space

The seventeenth shuttle mission (STS 51-B, flown in April–May 1985) was also known as Spacelab 3. Besides the commander and pilot, the seven-member crew included two M.D.s and three Ph.D.s responsible for doing scientific research. On the schedule for their week in space: fifteen investigations in life and materials science, fluid physics, atmospheric science, and auroral and astronomical observations.

Scientists outside NASA had developed the experiments and trained this crew to conduct them. One of the onboard scientists was flying specifically to operate an experiment facility that he had spent a decade developing; he was one of the first scientists to go into space to do his own research. Also aboard were twenty-four white rats and two squirrel monkeys, flying to test an animal habitat that would admit rodents and small primates to routine spaceflight for laboratory research.

Comments overheard during the Spacelab 3 mission hint at some of the reasons that NASA reframed the meaning of human spaceflight as scientific research.

"It's working! It's working," payload specialist Taylor Wang announced when his instrument finally began to operate after a three-day troubleshooting and rewiring effort.

"I'm concerned about his water and food intake," said mission specialist Bill Thornton, a physician, tending a small primate having symptoms of space sickness.

"It will take years to analyze all the data," and "we've collected more data than our wildest dreams," pleased scientists told reporters near the end of the mission.

Not since the Skylab missions in 1973–74 had American scientists had access to a laboratory in space. As demonstrated on the Spacelab 1 mission in 1983, the space shuttle offered them the capability to do hands-on research in space, to tinker with balky equipment, and to intervene in experiments just as they did on earth. Research began moving to the fore in human spaceflight.

No sooner had the space shuttle become familiar as a space truck that successfully delivered and retrieved satellites than NASA began to reframe the shuttle in the early 1980s as a laboratory. Billed as a multipurpose vehicle, the shuttle had carried small experiments on each of its early missions, but its real utility for research would be realized when it was equipped with a pressurized laboratory. A lab in space connoted pioneering research conducted by a crew of scientists. This purpose of human spaceflight was the intellectual complement to the muscular business of the satellite delivery missions. The useful, practical shuttle would become an instrument of knowledge and discovery.

As always, NASA developed a graphic logo to brand this program. The Spacelab logo depicted people working in a laboratory in space. Two stylized human figures inside the cross-sectional shape of a Spacelab module occupied the center of the circular design. A shuttle shape with objects in the cargo bay appeared to be in orbit on the border of the circle. It may not have been fine art, but the logo captured the essence of Spacelab's identity and purpose (fig. 4.1).

How and why did NASA begin shifting gears so soon in the shuttle era? What was NASA trying to accomplish, and did it influence the popular perception of spaceflight as it was characterized in the media and popular culture at the time? What do the rhetoric and imagery of this period reveal about the meaning of spaceflight and the role of humans in space? Hailed as "the flying Ph.D.s," shuttle-era astronauts brought to life the idea of "manned space science" as a highly collaborative and interactive exploration of research fields that might be productively pursued in orbit.[1]

NASA's first priority for the 1980s was to make the space shuttle operational to meet the promise of routine, economical spaceflight, and the agency directed a large share of its energy to that goal. But senior management believed it was important to have a vision with a major new goal on the horizon, and the best

Fig. 4.1. Figures of people are central in the Spacelab logo, set inside a shape that depicts the angled interior and circular exterior of the laboratory module flown inside the space shuttle. NASA (S76-29665).

opportunity to promote it would be while the new shuttle was in the spotlight. Immediately after the first shuttle flight in April 1981, NASA administrator James Beggs and deputy administrator Hans Mark agreed to make NASA's next major initiative a space station. They launched a carefully orchestrated effort to build the case for a permanent human presence in space and to campaign for President Ronald Reagan's approval to develop a space station.[2]

Mark reported that their strategy for persuading the administration to approve the space station as a new initiative consisted of two tactics: focusing their efforts of persuasion on the president and his immediate staff, and using "the very favorable public reaction to the space shuttle missions to draw the president's attention to the possibilities of what could be done of political value with the space program." They assumed that if they could convince the president that a space station was a good idea, he would issue a Kennedy-style mandate that would muster the requisite political support and resources. President Reagan ultimately did that in 1984 when he authorized the development of a space station within a decade, but the rest of the pieces did not readily fall into place.[3]

NASA's space station planners envisioned a large multipurpose facility that would be a combined research laboratory, satellite servicing center, and staging base for expeditions. However, the agency faced a significant early obstacle to consensus on a space station: lack of interest, or only lukewarm interest, in the scientific community. Thus, the scientific shuttle missions—the Spacelab missions—were crucial in NASA's effort to frame the case for a space station. Mark remembered that "many of us felt that it would be important to use the shuttle to the utmost to demonstrate what new capabilities the presence of people in space would open up for us. The first flight of Spacelab scheduled for November 1983 would be such an opportunity." Spacelab missions would test those capabilities and preview the science that a space station eventually would host. For this reason, NASA added space *lab* to space *truck* in its shuttle rhetoric.[4]

Spacelab developed in tandem with the space shuttle during the 1970s as a way to make space readily accessible to researchers. It was a suite of elements carried in the payload bay—an enclosed laboratory module where scientists could carry out investigations in various research disciplines, and a set of open platforms for telescopes and other instruments needing direct exposure to space. The European Space Agency (ESA) developed Spacelab under a cooperative agreement with NASA. Spacelab marked Europe's entrée into human spaceflight through a technical and managerial apprenticeship in developing flight hardware, and it was their ticket to flight opportunities on the space shuttle. The circular logo of the multinational European Space Agency branded the hardware and, appearing with the NASA logo, enhanced the prestige of this new entrant into human spaceflight. The words "ESA NASA SPACELAB," and the flags of the participant nations that often appeared on the Spacelab logo, recognized Spacelab as a cooperative endeavor.[5]

Spacelab expanded the capabilities of the shuttle from a delivery truck into a versatile orbital laboratory. In fact, it turned the shuttle itself into a temporary space station. Flying for one to two weeks at a time with four to six scientists aboard for two-shift, around-the-clock operations, Spacelab could accommodate tens of experiments and a variety of research equipment, from small instruments to furnaces, a glovebox, animal habitats, and devices large enough to seat a human. Investigators from around the world set up posts in the operations support centers in Houston, Texas, or Huntsville, Alabama, where they could stay in touch with the crew during the mission to troubleshoot problems or monitor experiments in progress, almost as if they were in their own labs. This unprecedented direct contact between researchers on the ground and in space gave Spacelab considerable appeal and raised expectations for high-quality research results.[6]

The idea of Spacelab had grown from studies in the United States and in Europe dating from 1969. With fading prospects for a space station successor to the Apollo program, planners considered whether a short-duration "sortie lab" would be a useful start to explore how scientific research could benefit from humans in space. NASA's director of the Spacelab program from the outset recalled the impetus for the idea of a manned laboratory: "Everyone looked for ways to make the shuttle capable of accomplishing many of the scientific objectives of the Space Station, until such time as the Station would finally receive a go-ahead."[7]

Years before the shuttle or Spacelab was in service, Wernher von Braun, who had moved from Marshall Space Flight Center to NASA Headquarters to guide future planning, thought the agency should do more to promote the shuttle as an opportunity for scientists to participate in conducting their own research in space. In a 1971 memo, he wrote, "This potential benefit clearly needs to be described in some detail to members of the scientific community who neither understand what role they would play . . . nor how such a mode of operation would benefit them." He urged the manager in charge of cultivating the scientific community to develop "good visual material"—specifically, fifteen to twenty pictorial, artistic, textual graphics—to reach those who were uninformed or skeptical. "I think there is a very good story to be told" to create an awareness in the scientific community of the potential benefit of the shuttle.[8]

Most researchers accustomed to putting instruments on high-altitude balloons, sounding rockets, or satellites were wary of the greater expense and complexity of "man-rating" their experiments to be part of human spaceflight missions. Others were curious but cautious about possibilities for "manned research" on shuttle sortie missions lasting up to a month. Researchers in life and biomedical sciences were optimistic about using a laboratory in space for reduced-gravity experiments that could not be done on earth. In a series of workshops and reports in 1971–73, scientists formed various discipline groups to explore how the shuttle might be utilized for their research and to scope out what kinds of facilities and capabilities would be needed. Representatives from the observational sciences of astronomy, earth and atmospheric science, and oceanography, plus two laboratory sciences (life and materials sciences that embraced several subdisciplines) and two technology disciplines found that a surprising variety of research potentially could be accomplished by scientists conducting experiments on the shuttle. The results of these efforts became the scientific basis for the development of Spacelab.[9]

Still, a "manned research lab" in space was not an immediately appealing concept to the scientific community in general. NASA would have to convince researchers otherwise, because the space shuttle needed customers and frequent flights to meet the goal of routine, economical access to space. To build interest, NASA engaged in outreach and marketing to educate and recruit the scientific community. A shuttle "utilization" office worked to make connections with universities and other research organizations, present talks at professional conferences, and distribute information about opportunities on the shuttle and Spacelab. NASA scientists involved in flight experiments also served as advocates for these capabilities.[10]

Scientist to scientist, the main selling point for Spacelab was the opportunity to do laboratory experiments in space interactively, similar to experimentation on earth. Experiments could be run multiple times with different settings or samples to obtain good data, and they could be modified and reflown on other missions. The obvious difference was that scientific investigators would work through surrogates, crewmembers who were trained to operate and optimize each experiment. In some cases, investigators would be able to join a mission crew to tend to their own research under a contractual arrangement that NASA offered. The first person to benefit from such an arrangement was a McDonnell Douglas chief test engineer in charge of developing a new technology that had commercial promise. The device flew on four shuttle missions operated by crewmembers whom he had trained; then, NASA certified him to fly with the device on three subsequent missions. The ability to train an astronaut crew and coordinate with them during a mission to carry out iterative experiments became a persuasive factor for thousands of investigators who proposed experiments for flight during the shuttle era.[11]

In time for the first Spacelab mission in 1983, NASA released several informational brochures for public consumption through its press sites and education centers. These publications introduced the concept of a versatile laboratory inside the space shuttle and outlined its advantages for scientific research. They also introduced the various science disciplines and explained why and how selected experiments would be conducted. These publicity materials painted a much different picture of human spaceflight than those for the satellite delivery missions. They emphasized people—the crew, investigators, and management team—and what could be accomplished by using the shuttle-Spacelab combination as a research center in orbit. Spacelab opened access to the unique environment of space for experiments that could not be done on earth. Human spaceflight enabled a new way to do laboratory research.[12]

When investigators whose experiments were chosen for the first Spacelab mission began to think about preparing a crew, NASA had eight scientist-astronauts left from the Apollo and Skylab era. In 1978 NASA selected twenty more scientists as mission specialists for the shuttle era. Mission specialists were heralded as a "new breed" of astronauts who would be responsible for everything happening on the shuttle except piloting. All held graduate degrees in science, engineering, or medicine, and they were well prepared to conduct research across the disciplines of space science and technology. Research in space was their domain.[13]

Imagine their reaction upon learning that another type of scientist—payload specialist—also would be on board for Spacelab missions. Payload specialists would be experts drawn from the investigator community, selected and trained to fly as guest astronauts to carry out scientific experiments. It appeared that NASA's professional astronauts would play a supporting role to these "non-astronauts," who represented and answered directly to the investigators for their particular Spacelab mission.

NASA introduced the payload specialist concept in press releases and Spacelab mission press kits as career scientists and engineers, men and women selected by their peers to conduct experiments on a particular mission. For each Spacelab mission, the investigators chose four candidates closely associated with one of the disciplines or experiments, who then cross-trained with all the investigators in their laboratories over a two- to four-year period. Every payload specialist had to become proficient on every experiment as well as integrated experiment operations to pass muster for their final selection: two would fly and two would be backups, working "on console" in the control center during the mission as the investigators' liaison to the crew or replacing a flight specialist if necessary. Payload specialists also received basic training from NASA on shuttle systems, procedures, and safety.

NASA's policy on payload specialists for Spacelab missions appeared in management guidelines developed in 1977. A payload specialist would be "a visiting specialist for a specific flight assignment generally selected by the investigators who will require his services," and these specialists "generally will be employees of the selected investigator(s) institution(s)." "Payload specialists will be the [investigators'] representatives, not NASA's," and thus NASA's involvement in the selection process was restricted. Christopher Kraft, director of the Johnson Space Center home of the astronauts, returned the draft guidelines with almost three pages of comments requesting clarification of the evidently overlapping roles of payload and mission specialists. Meanwhile, someone at NASA Headquarters was growing concerned about the selection process and confused roles. "We need all the good press we can get. . . . If either selection generates any bad press, i.e., protests of discrimination, etc., both will get the backwash."[14]

NASA's mission specialists argued that it was essential to integrate themselves into the Spacelab research teams and to participate in the intensive experiment training in order to be fully proficient members of a crew. Understandably, the "real" astronauts were concerned about being marginalized by non-astronauts. The very concept of mission specialists was still new, and these astronauts were eager to demonstrate mastery of their role. They asked whether they could be considered for payload specialist slots, and they received permission to apply, with the caveat that if selected they must be available for the considerable extra training. NASA forwarded names and resumes of interested, available members of the astronaut corps to the Spacelab 1 investigators for consideration, but they selected four payload specialist candidates from their own community. Despite the rebuff, NASA assigned two regular astronauts with equivalent mission-specific training and responsibilities to fly as mission specialists on each Spacelab flight.[15]

A year later, NASA administrator Robert Frosch directed that the science crew selection process be reevaluated, reportedly in response to Chris Kraft's continuing concerns about ambiguities in the roles of payload and mission specialists. He argued that mission specialists could fill the payload positions just as well, at lower cost, already having been carefully vetted as scientists and thoroughly trained in spaceflight. NASA briefly delayed the announcement of the Spacelab 1 payload specialists to hear him out. Frosch decided to let the payload crew selections for the Spacelab 1 and 2 missions stand, but he put the selection for Spacelab 3 on hold pending review of the process.[16]

Frosch was grilled during a 1979 congressional hearing about the agency's policy on payload specialists and mission specialists. NASA had recently clarified the roles of the two kinds of scientists, and set a policy that payload specialists would fly only when they were essential to the purposes of a mission "to perform unique tasks that require a specialized background." He noted that NASA's own mission specialists had sufficient expertise to conduct research on the shuttle, and it made sense to use the "permanent professional corps of astronauts" unless some highly specialized need arose.[17]

Several years later, NASA administrator James Beggs broadened the payload specialist policy to allow all shuttle customers and payload sponsors to select a representative to participate in their specific mission. Apart from health and training qualifications, NASA's only requirement was that the person "enhance the probability of mission success" by "supporting a legitimate scientific or development task." At some point, NASA issued a public affairs pamphlet describing the payload specialist role on the shuttle, and payload specialists flew on almost every science mission.[18]

Payload specialists were an attractive benefit for scientific users of the shuttle. Having a hand-picked, personally trained surrogate doing your work in space was almost as good as being there yourself, especially when you could talk to that orbital scientist almost at will to get the best performance from your experiment. For the payload specialists, it was a leap toward opening up spaceflight beyond the astronaut corps and a heady opportunity to study with first-rate scientists from around the world.

The tension between the roles of mission and payload specialists had another fallout—a debate over whether the professional astronauts should be broad generalists or experts in a narrow specialty. Most of those with Ph.D.s had entered the astronaut corps with already defined specialties, and some were actively continuing their own research. Yet to serve as a mission specialist, or perhaps to be chosen as a payload specialist, the astronaut would need to be an "effective generalist." The astronauts themselves and the space science and human spaceflight management weighed how to maintain individual expertise and foster cross-discipline expertise, how to balance research with training, and how to ensure the best-prepared, most proficient scientist-astronauts.[19]

The first Spacelab mission flew on *Columbia* in late 1983 as the ninth shuttle flight (STS-9). Unlike the four prior satellite delivery missions, Spacelab 1 had a crew of six working a twenty-four-hour, two-shift operation, and it had the longest duration to date—ten days. The payload bay held the twenty-three-foot-long laboratory module and a ten-foot-long exposed pallet loaded with observational instruments. This mission was designed to flight-test the Spacelab system, demonstrate its capabilities to support a variety of research, and attract international scientific involvement. For this proof-of-concept mission, most of the Spacelab hardware elements were included in the mission configuration, and the research roster listed at least seventy investigations for scientists from fourteen nations. Four members of the Spacelab 1 crew were Ph.D. scientists, two from the NASA astronaut corps and two selected by the investigators from their ranks to serve as payload specialists, one from the United States and the other from West Germany. The commander and pilot also assisted in some of the experiments.

NASA had tried "manned space science" during the Skylab missions of 1973–74, but Spacelab offered a more collaborative interaction between scientists in space and on earth. The science crew not only took care of routine research tasks—activating equipment, changing specimens, monitoring operations—but, more importantly, worked consultatively with investigators to seize opportunities, make fine adjustments, do repairs, and serve as test subjects for biomedical experiments. The science crew essentially became the eyes and hands, really the partners, of scientists on the ground. Having trained in their labs, the crew worked on their behalf and had in-flight conversations with them to achieve the best results for each investigation. This approach almost "put the investigator aboard" to optimize the science return of his or her experiment.[20]

Reusable like the shuttle, Spacelab also offered a broader and more flexible approach to research apparatus. More experiments and instruments could be installed, and they could be modified and reflown on successive flights. This ability to do iterative science, improving the hardware and procedures from mission to mission, was an important benefit of regular access to space and the investigators' close contact with the Spacelab crew.

Research on Spacelab 1 was organized into five disciplines: astronomy and solar physics, space plasma physics, atmospheric physics and earth observations, life sciences, and materials science. Experiments in life and materials sciences occupied the laboratory module, and most of the observational instruments resided on the exposed pallet. Crew members operated the laboratory experiments hands-on and remotely operated external instruments via computer commands. With such a multidisciplinary mix of investigations, there was naturally some competition for crew time and resources. Astronomical instruments had to be pointed by changing the shuttle's attitude, whereas materials science and life science research required stability to avoid disruption of delicate processes. Investigators had worked out most potential conflicts before the mission but met each day to negotiate changes and further compromises as the mission evolved.

The end-of-mission tally showed that almost all mission objectives had been met. Without surprise on a flight-test mission, a few anomalies had occurred in software and equipment, but only one of serious consequence. Investigators praised the crew's effectiveness and resourcefulness while on an exhausting timeline and found that having contact with the crew improved the outcome of their experiments. NASA officials declared the mission a successful merger of manned spaceflight and space science, and they judged Spacelab "a fantastic vehicle for performing science in space." Certainly there was an element of self-interest in these glowing reviews; scientists wanted more flight opportunities and NASA wanted more shuttle users, so positive reviews served both. Nevertheless, the Spacelab 1 mission validated the concept of "manned space science."[21]

While the first Spacelab mission understandably received widespread press coverage, perhaps the most thorough reporting was in the *New York Times.* Science correspondent John Noble Wilford filed daily reports from Florida, Texas, or California during the ten-day mission and both pre-mission and post-flight overviews. Over the course of a week, he accented each science discipline and many experiments, explaining what the crew did and why. His work offered a coherent account of human spaceflight for research, and it highlighted both the themes of NASA's public affairs material and his own assessment of science on the shuttle.[22]

Through Wilford's articles, the *New York Times* presented the public with a very positive account of the first Spacelab mission. Having greeted it as a "most ambitious test of the shuttles as vehicles for research," Wilford ended by quoting some of the principals. The NASA and European lead scientists assessed the mission as opening "tremendous opportunities for space sciences" and being "a great experience." One of the payload specialists reported that it was "a lot of fun" to work in the microgravity lab, while an investigator said the mission proved that nobody need ask "why a man is needed up there." Having worked through not only the planned research timeline but also the unexpected challenges, the teams in space and on the ground deemed the mission a success.

Indeed, the mission seemed to be an almost unqualified success until near the end. As the crew activated the lab for the first time, some to-be-expected nuisances occurred—an instrument wouldn't turn on, a short circuit or a software error occurred—generally things that were easily fixed by the crew. Later a few more consequential problems arose—failure of a high data rate recorder that resulted in lost experiment data, partial failure of an electron beam generator for one of the experiments—that were disappointing but not calamitous. The most troubling problems arose at the end when one, then two of the shuttle's general purpose computers and a navigation instrument, all used during reentry and landing, failed. Return was delayed eight hours while Mission Control reconfigured the backup computers to take over. *Columbia*'s return then appeared to be uneventful until the landing team later noticed signs of a small explosion and fire in the engine compartment. These final-hour orbiter malfunctions—none originating in Spacelab—triggered an extensive post-flight analysis.

Wilford's reports gave a sense of how the investigators and crew were able to work together as scientists. One day he narrated some of the conversation between payload specialist Ulf Merbold and an investigator in Houston as they jointly carried out one of his experiments almost as if they were side by side. He also gave insight into the investigators' input into mission management decisions, such as determining a favorable launch date or deciding whether to extend the mission for another day. The group had to balance sometimes competing interests and reach consensus on such matters as how to reallocate crew time if an instrument needed diagnosis and repair or if it failed. Extra time for one experiment meant less time for another, and turning off one instrument opened up power and crew time resources for others. With advocates for seventy experiments, negotiations could be complex, as the group had to make tradeoffs to gain a successful mission result for everyone. Working together in this spirit, the investigator group successfully lobbied mission management for an extra day in orbit to complete more science.

Wilford also noted the very characteristics NASA promoted: the international cooperation and partnership with Europe that Spacelab represented, the new style of around-the-clock crew activity, and a crew strongly tilted toward science. He quoted mission commander John Young's wry observation, "For the first time in space flight, the doctors outnumber the pilots four to two. It's going to be a very interesting and unusual mission."

Beyond reporting the mission events, the *New York Times* also presented the bigger picture. Even as NASA was somewhat coy about its intentions, Wilford realized the motive for this demonstration of what can be done in a laboratory in space. He stated that Spacelab was "a precursor of a permanent manned space station," an example to persuade others of the value of a large space station, permanently occupied by scientists.

Hans Mark, one of NASA's chief strategists for a space station, spent a mission day with Spacelab 1 investigators in the operations control center in Houston, watching their interactions with the science crew who were performing their experiments. It was then that he recognized the significance of Spacelab: "I had not realized earlier that Spacelab and its operation would be a natural precursor to the space station and that we should consciously begin to look at [it] that way." Others also recognized that Spacelab might well foretell the value, and even the appearance, of a future space station. Might Spacelab elements become the building blocks of an orbital research station?[23]

The next Spacelab mission to fly, Spacelab 3 (STS 51-B in 1985), moved ahead of Spacelab 2 because of a hardware delay for the second flight-test mission. Although it was multidisciplinary like Spacelab 1, with just fifteen investigations Spacelab 3 aimed more at depth than breadth of research. It also carried more than twenty small laboratory research animals to test enclosed habitats and procedures for their care. To accommodate the mission's emphasis on microgravity research in life sciences and materials science, the pilots placed the orbiter *Challenger* in a stable tail-to-earth and wing-forward "gravity gradient" attitude that would not require in-flight thruster firings, thus maintaining a quiet, no-vibration ride for the delicate experiments. As this first "operational" laboratory mission unfolded, it inadvertently demonstrated the value of scientists to deal with unanticipated problems in space, and conversely demonstrated some unanticipated problems of dealing with people in space.

Problems began before launch when the two squirrel monkeys designated for the mission were found to have a herpes virus that could be transmitted to the crew, and a new pair had to be prepared for flight. In orbit, the allegedly self-contained onboard animal habitat proved not quite as airtight as expected, allowing food and waste particles to escape into the cabin. The latch on the scientific airlock, through which a camera was to be deployed, jammed and rendered the camera useless—the only experiment to get no data during the mission. An instrument to measure chemicals in the atmosphere leaked its coolant and shut down. A sophisticated fluid physics facility would not turn on. A couple of rats and one of the monkeys would not eat or take water, apparently suffering from space sickness.

The ways in which the crew and the scientists on the ground handled these situations revealed both the merits and challenges of human spaceflight in the service of research. Besides the commander and pilot, the crew included three NASA mission specialists—two physicians and one physicist—and two payload specialists who were Ph.D. scientists representing the investigators whose experiments were installed in the lab. One of them, Taylor Wang of the Jet Propulsion Laboratory, was in charge of his own experiment apparatus for studying the dynamic behavior of fluid drops being manipulated by acoustic fields in microgravity. The other payload specialist, Lodewijk van den Berg, a chemical engineer from EG&G Corporation, was an expert in crystal formation for high-technology uses. One of the NASA scientists, Bill Thornton, M.D., served as primary caretaker and veterinarian for the animals.[24]

A problem with the animal enclosures became evident almost immediately when Thornton observed that the rats' dry food bars were disintegrating into dust. As the little animals tumbled around in their enclosed cages, they apparently kept the air stirred up beyond the capacity of the habitat's air filtering system. When Thornton went through the normal routine of checking their food supply and waste output, a cloud of dusty particles escaped, quickly dispersed through the lab, and made its way into the crew cabin.

Scientist Thornton went to work on finding a solution, but mission commander Robert Overmyer was clearly annoyed by this contamination. Mission Control advised the crew to don surgical masks for any activity involving the animal enclosures, and this unsanitary breach factored into NASA's decision to ground the habitat facility for design improvements after the mission. Meanwhile, Thornton worked with the animal facility team on the ground to mitigate the problem and with other crewmembers used the onboard vacuum to clean up the mess. What was supposed to have been a fairly straightforward animal monitoring task became a time-consuming housekeeping chore. The lesson from this incident: more effort was needed before animals and people could share space on the shuttle.

A more positive outcome of human presence resulted from the crew's tending of the animals. As physicians with extra veterinary training, Bill Thornton and Norman Thagard were responsible for monitoring the animals' behavior and condition. They intervened to hand-water some of the rats and hand-feed one of the monkeys until the animals learned how to use the dispensers. Thornton gave special attention to a lethargic monkey that seemed to have space motion sickness, and he happily reported that it revived after the human contact. During a televised tour around the lab a few days later, the animals appeared to be healthy and acclimated to spaceflight. The crew returned with all animals alive and had complete data for the investigator teams.

Then and thereafter, whenever animals flew, the media reported on them, sometimes with a sense of humor. In reality, use of animals in research was a sensitive topic that occasionally prompted protests against the missions, and NASA released very few images of the spacefaring animals into the public domain. Editorial cartoonists had fun with the animal presence in space, especially the Spacelab 3 "mishap," and with the idea of the shuttle as a modern-day Noah's ark (fig. 4.2).[25]

The most trying problem that ultimately yielded to human effort involved a $3.5 million facility for fluid dynamics research, designed and meant to be operated by guest payload specialist Taylor Wang. When it would not power up, the entire week of experiments was in jeopardy. Wang worked relentlessly to find the problem and ultimately isolated a short circuit deep inside the refrigerator-size device, rewired it, and salvaged three days of operation. Spacelab had touted the benefits of entrusting experiments to very capable onboard scientists, and this episode affirmed that benefit. His successful effort took a toll, though. Wang worked almost around the clock, declining to pause for meals or sleep, to fix the device and make up for lost operational time. Besides putting his own health in jeopardy, his dedication to this primary task meant that other crew members had to pick up his secondary tasks. They did so willingly to support an important research opportunity, but the incident did raise issues about balancing crew time and responsibilities.

The list of mishaps grew during the mission, but so did the list of successes. Despite the problems involving the animals, the airborne waste particles that forced the crew to wear masks, the investigation that was completely thwarted when the scientific airlock would not open so a camera could not be deployed for astronomical imaging, an atmospheric chemistry monitor beyond the crew's reach in the payload bay that leaked and shut down, and a small satellite that could not be launched because its batteries failed, the crew still managed to gather data. Two crew members, for example, jerry-rigged a cable to revive

Fig. 4.2. Flying animals on the shuttle for serious research in gravitational biology also brought moments of annoyance and humor. Published in the *Cleveland Plain Dealer;* courtesy of Ray Osrin Editorial Cartoon Collection, Michael Schwartz Library, Cleveland State University.

another instrument threatened with demise. And although half of the fifteen investigations needed more than routine crew involvement to meet their objectives, the other half—especially the crystal growth experiments and auroral imaging carried out by Lodewijk van den Berg and Don Lind—exceeded expectations. The crew and most of the investigators ended the mission well satisfied with their scientific productivity despite equipment problems.

News coverage focused on two aspects of the Spacelab 3 mission: animals and problems. What reporter could resist the human interest angles of a space-sick monkey, errant food and fecal matter, and broken equipment? Editorial cartoonists had some fun with the "monkey business." Yet during the week-long mission, coverage trended from an accent on problems to an accent on problem-solving. Early in the week, reports highlighted the various glitches and failures plaguing the crew. By the end of the mission, the crew, especially Bill Thornton and Taylor Wang, received notice for their perseverance in resolving problems. Reporters seemed to be probing beyond the superficial events to understand what the mission really entailed.[26]

The Spacelab 3 mission put the concept of "manned space science" to the test. The crew, in concert with investigator teams on the ground, managed to achieve mission success despite a rocky start. Without their hands-on intervention and resourcefulness, several experiments would have failed to operate as planned. By the end of the mission, the investigative teams tallied up their research objectives and reported that they had achieved about 90 percent of what they had hoped for. The lesson for scientists hoping to use the shuttle for research was that despite careful engineering and testing on earth, unforeseen problems would occur that would tax everyone's resourcefulness. The crew and the team on the ground would have to resolve those problems, but they might not be able to do so satisfactorily.

After an intervening satellite delivery mission, the next Spacelab mission began within two months of the previous one. Spacelab 2 (STS 51-F in 1985) was actually the second test flight of the new research facility. Unlike Spacelab 1 and Spacelab 3, this mission did not use the laboratory module. Instead, the shuttle became an orbital observatory, its payload bay filled with three platforms or pallets of instruments for observing the sun, stars, space, and the upper atmosphere, plus a small satellite to investigate the environment around the shuttle. Five scientists operated these instruments from the crew cabin without direct hands-on contact, and the two pilots carried out many complex maneuvers to achieve the desired observations. The crew also handled life science experiments mounted in their mid-deck living area. Like the previous two Spacelab missions, this one had a twenty-four-hour, two-shift schedule and a long list of mission objectives for its thirteen investigations.

Serious problems hampered this mission from the start. Its first launch attempt ended with a heart-stopping automatic shutdown just two seconds before ignition of the unstoppable solid rocket boosters. In the next attempt two weeks later, the shuttle launched uneventfully until one of the three main engines shut down early, causing the vehicle to enter an orbit more than fifty miles lower than desired. The launch delay had moved the mission into the full phase of the moon with less favorable darkness conditions than desirable for some of the instruments. One of the computers that would control science instruments failed. The pointing mount designed to keep instruments precisely fixed on their astronomical targets while the shuttle maneuvered did not work at all until very late in the mission. As the mission began, it seemed doomed to partial failure.[27]

A week later, the mission team and the media proclaimed Spacelab 2 a success. The reason: the persistence of the crew, investigators, and mission management in replanning the entire mission "on the fly." Adapting to the lower altitude and moonlight by revamping the observation target list, tackling the pointing system by rewriting its control software, developing other "workarounds," and adding an extra day in orbit, the crew and mission team compensated for many of the problems and salvaged the mission. Before the mission, commander Gordon Fullerton had joked about this crew having more degrees than any other: if problems arise, "we'll just overwhelm them with education." His words could not have been more prophetic.[28]

By the final press conference, the mission scientist reported that most of the mission objectives had been met, although not exactly as originally planned. As a test flight, Spacelab 2 was a shakedown cruise for the instrument pointing system, which performed less reliably and precisely than expected. The mission also yielded a satisfactory amount of raw scientific data. Spacelab 2, like Spacelab 3, demonstrated the value of onboard experts to diagnose, repair, or modify scientific apparatus as they might in a laboratory on the ground. The ability to respond to unforeseen problems and to work collaboratively with investigator teams "in real time" proved the workable union of space science and manned spaceflight.

The first three Spacelab missions gave scientists and observers a basis for an initial assessment of "manned space science" aboard the shuttle. Was it a good environment for doing research? Was the interaction between crew and investigators productive? Were all the disciplines equally at home on the shuttle, or were some better accommodated than others? How well did the payload specialists integrate into the NASA crews? Could the shuttle really be well used as a short-term laboratory? Could credible research be done during a one- to two-week shuttle mission?

Journalists provided the first draft of this assessment via mission-in-progress news coverage. Various reporters watched keenly to address these questions and test the research rationale that NASA had offered. The news media played an active role in framing the purpose of human spaceflight as research, and there was general optimism on several points.[29]

Versatile scientist-astronauts. The early missions demonstrated that onboard scientists could conduct high-order research and also be invaluable for lower-order troubleshooting and repairs as needed. However, reporters heard notes of frustration and impatience when the science crews had to spend inordinate time on basic chores.[30]

Interactive science. Communication between scientists in orbit and on the ground contributed to mission success, as an aid to problem solving and as a way to milk more data from the experiments. But it could also become testy if the crew began to feel overtasked or micromanaged. It was evident that amicable relations could be strained if the workload became stressful.[31]

Perspective on problems. With increased Spacelab mission experience, it became easier for journalists to distinguish between nuisances and genuine problems, and to understand better the nature of experimental science. Experiments and instruments sometimes had to be coaxed into operation, and such adjustments were a normal part of research. With this insight into the scale of problems, alarmist headlines became less frequent. At mission end, most scientists could claim very high tallies of mission objectives met and data collected despite problems encountered along the way.

Research significance. NASA's pre-mission science briefings, science summary briefings, and printed materials about the science agenda for Spacelab missions served their purpose as journalists took pains to explain how and why experiments were conducted. Although major discoveries probably would not occur during a mission and data analysis might take years, reporters tried to convey the longer-range significance of onboard research. They introduced areas of science probably unfamiliar to many readers, such as crystal growth, vestibular function, and atmospheric chemistry, so the public could better understand the relevance of research on the shuttle.

Research productivity. News headlines conveyed scientists' elation with the amount of data collected during the seven- to ten-day Spacelab missions. Although results were preliminary, investigators and journalists alike were eager to report possible discoveries. Mission managers tallied how many millions of bits of data or hundreds of images had been collected to convey the productivity of research in space, and investigators released "quick-look" results at mission's end to hint at what might come through postflight study. On this basis, Spacelab seemed to be living up to its promise.[32]

Spacelab significance. Journalists heard and publicized one message more openly than NASA did: that Spacelab was a precursor to a space station. NASA had not yet published that link directly, but people were talking about it, including Spacelab 3 mission manager Joseph Cremin, whose remarks were quoted in the press: "We look at Spacelab as being the springboard to a space station," and the experiments are "natural predecessors" to research on a space station.[33]

Routine spaceflight. NASA's Spacelab 3 press kit opened with the statement that this mission would "usher in an era of routine flights for Spacelab." Spacelab 2 followed two months later, marking the nineteenth shuttle mission to date, and the eighth for *Challenger* since its introduction in 1983. Journalists noted that this quick pace, with shrinking intervals between flights, looked promising for a mission-a-month flight rate. The space transportation system was maturing.

NASA had consciously put research into the routine spaceflight frame to appeal to the scientific community, and the media validated that concept. At the same time, the press also reported that NASA felt pressure to show the shuttle's reliability by adhering to a tight launch schedule. Yet the problems that affected *Columbia*'s return on Spacelab 1 and *Challenger*'s launches on two Spacelab missions suggested that the shuttle might not yet be reliable.[34]

The scientific community also made its own early assessments of Spacelab's utility and the value of human spaceflight for "manned space science." The first three Spacelab missions demonstrated the shuttle's use as a laboratory for multidisciplinary research and interactive scientific collaboration between investigators on earth and in space. These missions also demonstrated the complexity of sharing time and resources equitably among multiple investigations and the resourcefulness of the crew and ground teams in resolving unanticipated problems. Investigators judged these missions successful in meeting their specific objectives. But measuring their success in reframing human spaceflight as research, as NASA sought to do, was more difficult. As science was the reason for the missions, the broader standard for assessing their success must also be science—its quality and quantity and the value added by conducting research in space. NASA, the European Space Agency, the investigators, and the scientific press began to make these assessments during the period from the first Spacelab mission in 1983 through the third mission in 1985.

The journal *Science* raised the critical question just before the Spacelab 1 launch: "How useful will the shuttle really be for science?" It gave a more nuanced perspective on the scientific utility of Spacelab and the shuttle by revealing some behind-the-scenes dissatisfaction and reduced expectations. The main problems were fewer flights on the horizon and the cost of human spaceflight. Preparing a "man-rated" instrument for Spacelab cost three to five times more than developing one to launch on a rocket (though less than developing a satellite), and the lag time until flight could occupy a decade or more of a scientist's career. A NASA official noted, "The scientists don't trust us yet. . . . We have to give people the hope that they *can* refly." It was already evident that the shuttle was "hardly the ideal platform for high-precision science" because it vibrated and emitted vapors and exhausts that interfered with sensitive instruments. The shuttle-Spacelab would not be a perfect research environment, and for some experiments it would be a disappointing or unsuitable environment, but would it be good enough to do good science?[35]

Six months after flight, *Science* dedicated an issue to reporting the scientific results of the first Spacelab mission. Many of the investigators first published their results in this forum to reach a broader audience than more specialized journals. Noting that researchers are usually lucky to get any results at all from a new experiment, the editor remarked that "the achievements thus far are a good omen for further successes."[36]

Two years later, upon completion of the first three Spacelab missions in 1985, the picture was somewhat cloudier. Scientists were getting most of the data they wanted, sometimes more, but things did not always work as planned. Equipment breakdowns had been the biggest challenge, taking crew time to resolve instead of operate, but the ability of crews to work with the teams on the ground had turned equipment problems into a success story.

Was the science done in orbit pioneering or elementary, interesting or boring? Opinion among scientists varied. Those whose experiments were chosen for flight on the shuttle had a more positive view than those who did not participate, and scientists from some disciplines found Spacelab more hospitable than others. Some in the scientific community were disgruntled because the cost of the shuttle program had eroded funding for other research unrelated to the shuttle or Spacelab. Astronomers generally thought the shuttle would be more valuable for delivering and servicing large payloads, like space telescopes, than for carrying observatory platforms, due to its vibration, electrical and magnetic fields, and propellant contaminants. The one- to two-week mission duration was too short for some investigations in all disciplines to collect enough meaningful data.[37]

Besides these limitations of the shuttle itself, the process for preparing investigations for space was longer and more complex than developing experiments for a lab on earth. Scientists had to pass through more gates to get an experiment into space—from initial competition to final cooperation. NASA released announcements of opportunity, and scientists responded with competitive proposals for peer review and selection within and among scientific disciplines. After that they faced a series of engineering milestones for instrument and operations design, testing, and certification to meet all of NASA's standards and then deliver the hardware and software ready for flight. They had to train the crews and their trainers to conduct the experiments, participate in simulations of joint experiment operations per the mission timeline, meet with other investigators to balance the competing needs of everyone's research, and finally carry out the actual mission from the payload control center. This commitment had taken nearly ten years for the early Spacelab scientists, a discouraging period of waiting to do their research.

Science journalist Dennis Overbye wrote that "a Spacelab mission is sort of a giant scientific commune," and "not everyone was happy with the communal experience." Some of the scientists were loners, focused on their own work and very competitive about getting it done. Spacelab missions required cooperation and consensus about sharing onboard resources and crew time. NASA sought to run the missions equitably for everyone's success, but the politics of science inevitably intersected in management decisions.[38]

After investigators for the first three Spacelab missions had presented their preliminary results within a year after flight, the chief mission scientists and mission managers commissioned an illustrated book-length report on the scientific harvest. *Science in Orbit: The Shuttle-Spacelab Experience* was NASA's first collation of research results, discipline by discipline, for the inaugural Spacelab missions and other science payloads in the first five years of shuttle flights. Spacelab 1 mission scientist C. R. Chappell, later the associate director for science at NASA's Marshall Space Flight Center, NASA's lead center for Spacelab, guided the project. He stressed the theme of "manned space science"—the union of space science and human spaceflight—as the essence of Spacelab and a more active personal participation in space research. Unlike the essentially passive mode of putting automated instruments on satellites or rockets and waiting for data, Spacelab investigators had the benefit of working with scientists in space during hands-on operations. Research results from the first three missions proved the value of that close collaboration, and papers began to appear in *Science, The Physiologist, Annals of Botany,* and other journals.[39]

This cumulative report validated the new meaning of human spaceflight as scientific research. Illustrated with scenes of the flight crew in the lab, scientists at work in the operations center, and data images from experiments, the report visually captured much of the experience of manned space science. With its focus on research results rather than experiment operations, it did not dwell on the few frustrations encountered during the missions, such as balky equipment and software. Nor did it address how the investigator groups negotiated their way through any tensions that arose during the missions. It was a positive assessment of this new opportunity for doing science in space, with a focus on the new knowledge and discoveries already attained. The final chapter looked ahead to the evolution of Spacelab research on a permanent space station.

The European Space Agency also produced a book, *Spacelab: Research in Orbit,* that narrated the Spacelab story from development through flight of the first mission. It took a similarly positive perspective on preliminary scientific results by discipline, with photographs that brought the mission to life. The Europeans reported that the team approach was very successful and the onboard scientists were invaluable. While lessons learned would lead to improvements on future missions, Spacelab 1 could only be judged a success. It validated the new concept of an international and multidisciplinary laboratory in space that "will surely lead eventually to a permanently manned Space Station."[40]

Not all assessments of Spacelab science were favorable. A vocal and influential group of scientists, primarily astronomers and physicists, had long criticized the shuttle program and human spaceflight for drawing funds away from other worthy scientific research. The use of the shuttle to support multidisciplinary research did little to assuage their criticism. Renowned space scientist James A. Van Allen, who opposed human spaceflight, was an oft-quoted spokesperson for the critics. He argued that almost all important science in space had been achieved with unmanned satellites and planetary probes, and that "the vaunted advantages of human crews in space vehicles, derived from a romantic vision of human space flight, have not been sufficiently subjected to critical assessment." He (among others) was not impressed with the rationale or record of manned space science.[41]

At the heart of such criticism lay a redefinition of space science. Long the bastion of astronomers, planetary scientists, and solar and space plasma physicists who worked with automated instruments, space science began to broaden in the shuttle era. The new-to-space biological and materials science disciplines were moving under the space science umbrella and gaining seats at the table of the Space Science Board of the National Academy of Sciences. To some degree, the old-guard space scientists resisted the encroachment of these new space scientists whose experimental methods required, and benefited from, human involvement.

After the first three missions, a new generation of Spacelab missions began. Known in-house as dedicated discipline labs, to the public they might have appeared as missions with a theme: German or Japanese, life sciences or microgravity sciences, astronomy or earth science. The multidisciplinary approach of the demonstration missions gave way to a more focused strategy of flying compatible instruments and investigations in a single discipline. Eighteen more Spacelab missions, from 1984 through 1998, became increasingly sophisticated laboratories or observatories for in-depth research in particular disciplines. The most specialized and coherent was Neurolab (STS-90 in 1998), tightly focused on studies of the brain and nervous system.

Of the disciplines that tried Spacelab, all stayed aboard for at least two more missions except space plasma physics and solar physics. During the 1990s, the missions were increasingly devoted to life sciences and materials (or microgravity) sciences—the two disciplines that most required the crew's involvement and that perhaps held the greatest promise for practical results and benefits on earth. With several observatory deliveries added to the count of Spacelab missions, the 1990s became a decade for science on the shuttle.

The fourth Spacelab mission (STS 61-A in 1985) belonged to the European Space Agency and Germany. Dubbed D-1 for Deutschland-1, to recognize the dominant role of Germany in developing Spacelab, this mission was granted to ESA in the international agreement. A follow-on Spacelab mission D-2 (STS-55) flew in 1993, and a dedicated Spacelab J mission for Japan (STS-47) flew in 1992. These missions honored internationalism in partnership with European nations and the Japanese Space Development Agency, also eager to participate in human spaceflight.

The D-1 mission marked the Europeans' full arrival in human spaceflight. They had three payload specialists on duty, responsibility for the entire payload, and management responsibility for science operations. Moreover, they worked in their own space operations center near Munich. Spacelab 1 had been their dress rehearsal, but now they were operating as an equal partner. The payload of seventy-five experiments emphasized materials science and life science. The second German mission had a more varied payload mix and two European payload specialists on board. For the Spacelab J mission, the fiftieth shuttle crew included a Japanese payload specialist, and Japan provided more than half of the experiments in materials science and life science.

The purpose of these dedicated missions was not simply to afford other nations an opportunity to send their own astronauts to do science in space; it was also to draw them into the fold as space station partners and prepare them for larger roles in a permanent research center in orbit. The Spacelab missions typified how scientific research would be conducted routinely on a space station, and they offered chances to test new experiment apparatus and procedures with time to refine them before sending them to a space station. While press coverage of these three missions in the United States was rather modest, it was huge in Europe and Japan, where Spacelab contributed to national pride and helped build enthusiasm for the space station.

The first of NASA's dedicated missions was Spacelab Life Sciences 1 in 1991 —the first shuttle crew to include three women and three medical doctors— followed by a second one in 1993, a Spacelab-*Mir* mission in 1995, a Life and Microgravity Spacelab mission in 1996, and Neurolab in 1998. NASA had a strong tradition of life sciences research rooted primarily in the Johnson Space Center in Texas and Ames Research Center in California. As home of the astronauts, Johnson's research interests lay primarily in understanding anything that might affect the health and wellness of the crew: space adaptation syndrome (space sickness), cardiovascular changes, metabolism, bone health, balance and orientation, and other issues of space medicine. The Ames Center's research interests included these but extended as well to more fundamental questions about the nature of living things in microgravity. Ames also had a strong tradition of animal research. NASA's life sciences missions brought together researchers from both centers and also from universities and other entities to use the microgravity environment on Spacelab to investigate questions about life sciences. For flight experiments seeking to understand the nature of physiological changes and recovery, investigators collected comprehensive data on the human, plant, and animal test subjects before, during, and after each mission.[42]

As it had done for the first three Spacelab missions in the early 1980s, NASA continued to publish a series of colorful booklets featuring the research plan for each mission. They framed the purpose of the missions and translated the experiments into a narrative intelligible to the press and other nonspecialists. Produced in magazine format, these booklets offered more than the basic information available in the mission press kits. Reflecting the missions, they organized the life sciences experiments by bodily systems and explained the reasons and methods for studying every organ or function of interest. They often included a detailed map of the laboratory layout and photographs of the crew and investigators engaged in research. In essence, these public information materials invited readers into the lab and gave a substantive-enough tutorial to foster basic understanding of the research. Spacelab scientists worked closely with the editorial team to frame the rationale for their research and NASA's life sciences objectives: to understand the relationship between life and gravity, to enable human exploration and habitation of space, and to bring benefits home to earth.[43]

The pool of research subjects on the life sciences missions included humans (the crew), rodents, jellyfish, crickets, snails, and fish, plus various plant seedlings, algae, bacteria, fertilized eggs, and other organisms. Humans had the most rigorous regimen, submitting themselves to blood draws, electroshock, implanted sensors, induced disorientation, and other discomforts for the sake of science. These experiments made for good "human interest" and "animal interest" stories in the press, and the media generally paid attention to the life sciences missions. NASA had so effectively articulated the research rationale that it was mentioned in almost every news report about these missions.[44]

The series of Spacelab life sciences missions demonstrated an important advantage of doing research on the shuttle—the opportunity for research to evolve from one mission to the next. The investigators' ability to retrieve their instruments, modify them if desired, improve procedures or software, and then fly them again made it possible to advance their space research much as they would in a laboratory on earth. A number of investigators flew their experiments more than once or twice for this evolutionary benefit. The life sciences missions also had extended duration, lasting nine to sixteen days, to allow for more time in microgravity and more data collection. While time on a shuttle mission was brief compared with indefinite time on a space station, the investigators found the two-week missions suitable for the types of experiments being flown.

The last of the Spacelab life sciences missions, Neurolab, flew in 1998. It was by far the most focused of any of the science missions, with all twenty-six investigations targeting the brain and nervous system. It brought together science partners from Europe, Canada, Japan, and various branches of the National Institutes of Health in a deliberate effort to expand involvement in space research to new communities. NASA saw this mission as a model for coordinated life sciences research on the International Space Station, whose assembly would start the next year. The Neurolab team gathered the results of their research into a single publication so that interested, scientifically literate generalists could easily learn about the mission. The sophisticated Neurolab mission signaled the maturity of human spaceflight and scientific research (plate 8).[45]

A relatively new discipline used the space shuttle most frequently. First called "materials science" before evolving into "microgravity science," this discipline blended chemistry and physics in the study of materials and the processes by which they form, take shape, and transform when subjected to physical forces—when molten metals solidify into alloys, for example, or crystals condense out of fluid or gas solutions. For this research field, a laboratory in microgravity truly offered a unique environment for doing experiments that cannot be done on earth, where gravity interferes with or masks subtle phenomena. The success of materials science research on the Spacelab 1, Spacelab 3, and first German missions led in the 1990s to a heavily subscribed series of two International Microgravity Labs, two U.S. Microgravity Labs, two U.S. Microgravity Payloads, and a Microgravity Science Lab, as well as the second German Spacelab and Japan's Spacelab missions.[46]

NASA's basic rationale for studying materials in space was straightforward: gravity influences all kinds of materials from their formation to the processes by which they are shaped. Alloys, ceramics, crystals, glass, metals, plastics, fluids, and even cells are subject to gravity as they are mixed, separated, heated, cooled, welded, and handled, usually with detrimental effects on their quality. Working with materials in space allowed researchers to better understand and reduce those defects, yielding a higher-quality product. This basic knowledge motive might also have practical applications for improving materials processing techniques on earth to yield, for example, higher quality crystals for use in semiconductors and electronic devices. NASA pushed this rationale a giant leap further by suggesting that materials science research could lead to factories in space to manufacture perfect materials or pharmaceutical breakthroughs that could lead to cures for diseases. These exaggerated claims earned a fair share of criticism, if not ridicule. In reality, many of the microgravity experiments were developmental, carried out to test techniques and equipment for doing research. Science in space was in its incipient phase.

As for the other Spacelab missions, NASA produced colorful, magazine-like booklets in collaboration with the materials science investigator teams. These booklets discussed each experiment's purpose and method and explained the laboratory facilities, taking the reader into Spacelab to see what would happen there. As materials science was somewhat less familiar than other disciplines, the teams took pains to ensure that their research plan would be intelligible to nonspecialists. NASA and the investigators distributed these publications widely to the media covering the missions and others, such as congressional and White House staff, who needed to understand the research agenda.[47]

The microgravity science missions posed a challenge to the astronauts' tradition of designing patches. How does one depict microgravity? The Spacelab 3 mission patch and a few others featured the shape of a crystal, in reference to various experiments in crystal growth, or symbols such as a Vitruvian man or DNA strand from the life science investigations. The word "microgravity" was too long for a compact patch design, so it was usually reduced to an acronym, such as IML for International Microgravity Lab. Finally, on two of the last research missions—the 1990 U.S. Microgravity Payload mission (STS-87) and the 2003 Microgravity Research mission (STS-107)—an apt symbol appeared: the Greek letter mu (μ), shorthand for "micro," and a lower case *g*, a common abbreviation for "gravity." Thus, μg, spoken as "micro-g," became a quick graphic emblem for microgravity, the unique environment for laboratory research in space (plate 9).

Until the shuttle came into service, materials science research was limited to seconds of microgravity in drop towers, rockets, or parabolic aircraft flights. The prospect of hours and days, up to two weeks, in weightlessness was a huge boon to materials scientists, who lined up eagerly to put their experiments on the shuttle. The spaciousness of Spacelab, both the laboratory module and the exposed platforms, offered another advantage—the ability to fly a large instrument or facility, such as a furnace comparable to one in a lab on earth. While some materials science experiments ran on automatic mode, many needed a human operator-observer to make fine adjustments for the desired result. Spacelab offered that as well and thus was a good fit for materials science research. Access to the shuttle and Spacelab prompted the Space Studies Board of the National Research Council to develop a research strategy and identify opportunities for microgravity studies in space—a clear sign of a maturing discipline. Altogether, more than 300 major materials science investigations were carried out on the shuttle, an indication of how welcome human spaceflight was as a research technique for a new discipline earning its gravitas.[48]

From the Spacelab 1 mission on, materials scientists reported satisfactory results from their research, and like the life scientists, they were able to do evolutionary research, building on lessons learned to improve their instruments and procedures from mission to mission. Their main complaint was vibration caused by shuttle thruster firings and crew activity. Materials science experiments are inordinately sensitive to disturbances, and the mission crews sought to maintain as quiet and stable an environment as possible. Nevertheless, the small jet blasts that kept the shuttle oriented properly, a crewmember working out on the treadmill in the crew cabin, or even a sneeze close to an experiment could ruin it. This issue required constant vigilance as the one inescapable problem of a human presence for this type of research.

Microgravity science came of age on the space shuttle. By the end of the 1990s, eleven microgravity science missions had been completed and the scientific literature on in-space materials research was burgeoning. NASA had found steady customers for Spacelab in life and materials sciences who also became the primary clients for a space station. During the 1990s, as scientific research dominated human spaceflight, NASA reconceived the space station. Planners stripped away some of the unrelated functions, such as a staging base for planetary missions or a factory for space manufacturing. The new space station would be what had already proven worthwhile—an orbital center for research in materials and life sciences. For many in NASA and the media, microgravity research on the shuttle paved the way to the space station.[49]

The first three Spacelab missions had included in the payload mix a variety of instruments for astronomy, solar science, space plasma physics, atmospheric chemistry and physics, and earth observations. Thereafter, those observational science disciplines also coalesced into dedicated Spacelab missions, flying without the laboratory module but using only the exposed platforms, or pallets, as instrument mounts. Two astronomy missions flew as Astro 1 (1990) and Astro 2 (1992), and three other missions flew as the Atmospheric Laboratory in Space (ATLAS) in 1992, 1993, and 1994. Each of these Spacelab missions recycled instruments from one flight to another, effectively extending the duration of each investigation and factoring in lessons learned from one flight to the next.[50]

These observatory missions differed from the laboratory missions in another key respect: the level of crew involvement. Most of the instruments ran in semiautomatic mode in the payload bay, requiring the crew to activate startup procedures or targeting commands, or to troubleshoot any problems that arose by working at computers inside the crew cabin. The crew had no direct contact with the instruments and did not actually have to see them through the window in order to control them. This research by remote control was not unlike what happens in observatories on the ground, and the investigators used a communications link to the crew to direct their research. Each Astro mission crew included two payload specialist astronomers who had connections to at least one of the telescopes, but in a controversial decision, NASA determined after the first ATLAS mission that the follow-on missions did not require payload specialists. The second and third ATLAS crews included only mission specialist astronauts.

NASA's booklets for these missions framed the main advantage of doing research on the shuttle: location, location, location. Being above the atmosphere improved everything, whether looking toward deep space, the sun, or down into the atmosphere. Of course, instruments on satellites were in routine use as research tools for these disciplines, so the rationale for human spaceflight had to be made. The Astro booklets made a compelling case for X-ray and ultraviolet astronomy but largely ignored the question of whether human spaceflight added value to this research. The Astro missions emphasized the telescopes, an innovative complement of four quite different instruments mounted together on a pointing system. The crew selected targets for observation, commanded the pointing system, and verified data acquisition.[51]

Similarly, the ATLAS missions filled a platform in the payload bay of the shuttle with out-of-reach instruments operated remotely by the crew from the orbiter cabin. For these missions, the pilots played an active role maneuvering the shuttle toward the earth or sun at different viewing angles required by the investigations. Most of the instruments in the ATLAS missions ran automatically with some crew tending, except the highly interactive space plasma physics experiments. The crew could fire an "electron gun" particle accelerator mounted in the payload bay to create artificial auroras and study the ionized environment near the shuttle.

Exploring the mysteries of deep space via the Astro instruments or examining relationships between earth's atmosphere and the sun via the ATLAS missions provided a strong enough research rationale that neither NASA nor the investigators had to promise practical benefits. Better understanding was enough. Yet because the instruments made highly sensitive measurements, news of "discoveries" occasionally emerged from these missions. NASA's booklets did not project a future for these disciplines on the space station. By the time Astro and ATLAS missions flew, the station had been stripped of observatory platforms in favor of laboratories dedicated to life and materials sciences. After their flights on the shuttle, these observational science disciplines moved back onto satellites to continue their research.

Neurolab (STS-90) in 1998 was the last scheduled Spacelab mission. Assembly of the International Space Station began the next year, and all missions thereafter were dedicated to that effort, except for planned servicing missions to the Hubble Space Telescope. However, it would be several years before the station would be able to support research. The scientific community and Congress, concerned about a long hiatus in science missions, appealed to NASA to schedule a few more. Congress added funds to the NASA budget for a science mission that became STS-107 (2003), a very successful sixteen-day research stint on *Columbia* until its shocking end minutes before landing.[52]

The laboratory for this Microgravity Research Mission was SPACEHAB, a commercial module similar to Spacelab that was loaded with experiments in life and materials sciences. The crew carried out investigations for more than seventy scientists, with about half of the experiments devoted to biology and biomedical research. The mission exemplified the benefits that had been anticipated decades earlier: around-the-clock uninterrupted hands-on research carried out by highly skilled astronauts working in close coordination with investigators on the ground. Fortunately, most of the data had been transmitted during the course of the mission, but almost all experiment samples were lost along with the seven crewmembers and orbiter *Columbia*. As a grand finale science mission, it left a legacy of accomplishment as well as tragedy.

Thereafter, attention turned toward fully outfitting and staffing the International Space Station to be a high-functioning research center. Many of the facilities and investigations on the station had a heritage that extended through Spacelab missions back to the early 1980s. Spacelab served as a bridge between Skylab, shuttle, and space station that enabled the evolution of human spaceflight in new directions. A large share of the credit for research becoming the primary justification for a human presence in space and the focus for the International Space Station belongs to Spacelab.

From the beginning, the scientific community was concerned about the performance of the space shuttle, Spacelab, and surrogates (astronauts and payload specialists) for scientific research. Would the facilities and crew prove adept enough, would the investigations be sophisticated enough, and would the science results be significant enough to warrant the high cost and risk of human spaceflight? As a metric for judging the success of human spaceflight for research, skeptics often cited the record of robotic solar system explorers and free-flying telescopes, and made the irrefutable argument that they produced more exciting and more significant results at far less cost than laboratory research in earth orbit. On that basis, James Van Allen raised the question, after the Spacelab missions as the space station was being built, "Is human spaceflight now obsolete?" He and others questioned the nation's commitment to human spaceflight, and criticized the shuttle era's contributions to science as modest, judged by the caliber of investigations and the lack of breakthrough discoveries. Others tried to use a metric of "cost per experiment" to assess whether human spaceflight was a good scientific value.[53]

One instance in particular brought this skepticism into high relief—John Glenn's return to space on the shuttle to conduct research on aging. Glenn had always wanted to fly in space again after completing his short Mercury mission in 1962 to become the first American in orbit. Years later, he urged NASA to do more to examine the similarities between aging and the body's changes in spaceflight. With support from the National Institutes of Health and gerontology researchers, he persuaded NASA administrator Daniel Goldin to add some peer-reviewed aging experiments to a science mission and put him on board as a seventy-seven-year-old specimen. Glenn flew as a heavily instrumented payload specialist on STS-95 in 1998; he was wired with sensors to monitor sleep and many other bodily functions, and he had a catheter implanted for frequent blood draws. By all accounts, he took the research seriously and served well on the crew.[54]

The Glenn return-to-flight mission, as it became known, had two effects. It created a surge of interest in the public and media, resulting in one of the largest crowds to witness a shuttle launch, and it opened a brief public conversation about science on the shuttle. Glenn made the cover and feature story of *TIME* magazine and appeared all over the news. Walter Cronkite returned to the air to anchor television coverage of the mission, and public interest resonated with memories of the heroic space-race years. The mission was a publicity bonanza for NASA on par with the 1983 launch of Sally Ride, America's first woman in space. The entire crew later received a public outreach award for increasing public awareness of the space program. However, much of the news coverage mentioned doubts about the merits of the aging research and speculation that it was a cover for rewarding an American hero and staunch advocate of human spaceflight with a shuttle flight. Glenn and NASA, of course, maintained that the research was legitimate and the flight was not a political favor or a publicity ploy. The *New York Times*' John Noble Wilford glorified the mission in nostalgia, while his editorial team dismissed it as a thin "veneer of scientific justification." As research, a sample of one person does not make a significant database, as Glenn admitted, but he noted that it was at least a start.[55]

Glenn's remark that in research you have to start somewhere might well apply to many of the experiments flown on the shuttle and Spacelab. They were opening new areas of research in search of intrinsic knowledge about how gravity and microgravity make a difference. Investigators expected not to make major discoveries but to learn how to do the research effectively in space. The scientists expected incremental advances rather than instant breakthroughs, and they cautiously announced "surprises" rather than "discoveries."[56]

It was customary for investigator teams to meet in conference a year after their missions to share preliminary results. Each investigation had a set of mission success criteria to be tabulated and reported to the mission manager and chief scientist before the shuttle landed. These immediate reports recorded quantitative metrics such as how many times the experiment ran or how many different samples or observations were attained, but in most cases only quick-look evaluations of data quality were possible; comprehensive data interpretation would require months. The anniversary meetings typically served as the first presentation of results among peers. Each mission's Investigator Working Group, under the guidance of the mission scientist, compiled a one-year post-flight report for the space agencies, and some took a further step by publishing the document for broader distribution.[57]

As evidence of the productivity of shuttle-based science, NASA occasionally compiled a summary of results or a bibliography of publications issuing from research on Spacelab missions. After the last Spacelab mission, a comprehensive compilation tallied almost 5,500 publications, about half in peer-reviewed journals and others mainly as book chapters, conference proceedings, and dissertations. Organized by disciplines, this multivolume report included introductory essays on the research questions and rationale for investigations conducted in space and then proceeded through an analysis of each experiment and resultant published contributions to the state of knowledge. It offered qualitative measures for the significance of research. Investigations identified as "landmark" experiments opened a significant new line of inquiry or answered a perplexing question that enabled the field to advance. Others achieved a better understanding of the techniques or technology needed to advance research. Some of the life sciences research led to new or revised textbooks. Very few investigations yielded little knowledge, typically due to equipment failures or unrelated problems that compromised the experiment. Microgravity materials science and life sciences research proved especially productive. This report concluded that all of the disciplines had benefited from research with human involvement on the shuttle and Spacelab missions, although the benefits varied in kind and degree.[58]

For the observational sciences, especially astronomy and solar physics, that used commandable instruments mounted outside the lab in the payload bay, the primary benefit was testing and calibration of new instruments and trials of observation techniques. The one- to two-week missions were too brief for sustained observations, and on multipurpose missions the time available for pointing the orbiter had to be negotiated with investigators whose experiments required stability. The crews were useful for activating and monitoring instruments and troubleshooting problems, but most of the observing programs were automated. These disciplines reaped practical benefits from the crewed missions, but investigators eventually migrated back to satellites as their research platform of choice. Crews were more actively involved in the earth observation and space plasma physics experiments. These disciplines reported significant scientific results as well as proof-testing of new instruments and experiment protocols.

Life science experiments depended heavily on direct personal involvement and thus were especially well suited for human spaceflight in both their objectives and their procedures. All of the subdisciplines in biology and biomedicine reported productive results from human-manipulated and crew-as-test-subject experiments. Although the one- to two-week missions were not long enough, and the experiment population not large enough, to generalize results broadly, scientists developed confidence in the trends they began to see in microgravity—how the human body and all forms of life change in space, complicating conceptions of what is normal. This research on the shuttle was foundational in planning research programs in space biomedicine and gravitational biology for the International Space Station, and possibly for even longer-duration spaceflight to Mars. It also resulted in the publication of several volumes of results in space biology, space physiology, and space life sciences, and development of an advanced research strategy.[59]

Materials research in microgravity was an impressive success, advancing the field and creating new subdisciplines. The onboard scientists' abilities to see features and processes normally masked by gravitational effects and to manipulate experiments contributed to all areas of research. These investigations yielded results in both theory and applications, and also became the basis for planning space station research. However, these advances were not without some dissension. In particular, a controversy arose over the protein crystal growth experiments flown on the shuttle and *Mir* when an evaluation published in *Nature* found the results unimpressive and more generally found microgravity research "of little importance."[60] Investigators argued otherwise.

In every area of inquiry, the shuttle-Spacelab crews—"the flying Ph.D.s"—performed as extensions of the investigators on the ground and vice versa. This constant collaboration proved to be effective on both sides of the relationship, with some participants remarking that it seemed like they were working in the lab together. John Glenn remarked before his return to space, "We've turned space into a new laboratory. . . . It's opened a whole new vista on what we can do in immunology, biotechnology, and many branches of science." Despite doubts about its value, the successes of human spaceflight as research set the stage for the next logical step for some of the sciences: a space station.[61]

For a variety of reasons, it proved difficult to visually capture the intensity and variety of research in orbit. The laboratory module interior was about twenty feet long and six feet wide, more like a tunnel than a photography studio. Experiments in racks lined the walls on both sides, most of them covered by control panels. Those requiring observation had a very small viewport or monitor, making photography difficult. It was hard to get a good shot of a crewmember doing something interesting in this spartan setup in such close quarters. Also, there was a good deal of clutter—cords, checklists, small tools, and other items suspended or attached everywhere. Regrettably, the photos of research activity on the space shuttle were not as entrancing as those of spacewalks or views of earth. Instead of photographs, videos provided the best imagery from these missions because they enabled onboard scientists to explain what they were doing. NASA occasionally made the crews accessible to the media, students, and special events through live downlink or recorded video sessions to build awareness of the shuttle as a laboratory and astronauts as scientists.[62]

The life sciences investigations tended to have visual interest when the crew handled plants, checked on animals, worked together as test subjects, or donned experiment apparatus (sensor harnesses, electrodes, and the like, which might be mistaken for torture devices). Biological research tended to capture attention because the subject matter was inherently familiar. The materials science missions offered little to see; almost everything of interest happened inside sealed chambers and furnaces. Spacelab missions did not include extravehicular activity, so there were no dramatic spacewalk images. To some extent the Spacelab missions proved visually disappointing, given that their main purpose was people doing science in space. Many images simply showed the crew setting up equipment, checking checklists, peering at a monitor or viewport, and doing similar ordinary tasks. The in-flight photos were less interesting for public use than for the investigators.

By contrast, the data images were spectacular. Images of crystals formed in space rose to the level of art. Electron-microscope images of crystalline and metallic structures offered glimpses of perfection. Images of seedlings with tiny roots extending upward and sideways said more about microgravity than words alone. Floating spheres of fire from combustion experiments were wondrous. To those who read charts and graphs well, the spectral data from astronomical and atmospheric instruments were fascinating.

Newspapers and magazines found themselves somewhat at a loss for images during the Spacelab missions because interesting data images might not become available until after the news cycle. NASA and the investigators tried to convey the intrinsic interest and even wonder of their work with tantalizing images in the booklets circulated before the missions. The retrospective *Science in Orbit* published in 1988 brought together images and results from the first five years of shuttle missions, but neither NASA nor anyone else has yet done justice, in words or images, to the entire experience of scientific research on the shuttle.

Although they had sparse imagery to use, reporters generally did a good job covering the shuttle's research missions, featuring many of the experiments and profiling the scientist crewmembers. Both the *New York Times* and *Washington Post* headlined the main thrust of these missions—exploring how the body adjusts or how zero gravity affects developing plants and animals, for example—and they reported surprising early findings. In almost every instance, news articles touched on the research rationale for human spaceflight and conveyed the message that it was yielding new knowledge that might lead to tangible benefits on earth.

Spacelab is perhaps the best single icon for focusing human spaceflight on research in the 1980s and 1990s—the tangible realization of the idea that space offered a unique environment for research and that working there was a suitable occupation for scientists. From the outside, the long canister wrapped in white thermal blankets was impressive only in size, but the interior—a versatile, well-equipped laboratory where scientists performed hundreds of experiments—made it historically significant. Recognizing Spacelab as the iconic object for human spaceflight for the purpose of research, the Smithsonian National Air and Space Museum acquired one of the two flown laboratory modules. It flew in the shuttle nine times, from Spacelab 1 to the Microgravity Science Lab missions, before being retired. The Spacelab missions and other research on the shuttle contributed to knowledge in a spectrum of sciences and technologies. This research on the shuttle also set the stage for a permanent research center (space station) as the next logical step for a human presence in space.[63]

Chapter 5 Space Station: Campaigning for a Permanent Human Presence in Space

In March 1982, with only two flights of the space shuttle completed, NASA began a new campaign. A senior agency executive held an interview with the *New York Times'* chief science writer, John Noble Wilford, about a new project. His article, "The Shuttle's Future: NASA Looks Toward Space Stations," discussed the challenging new goal to establish a permanent human presence in space. The NASA official then sent him a three-page position paper by NASA administrator James M. Beggs, "A Space Station for America." This document made the case for "a permanent presence" in earth orbit, listed a variety of roles and operations for a space station, and described two possible designs then under consideration. The NASA document must have informed the interview, because the *Times* article was quite similar in content and rhetoric. Both pieces sounded the theme that would guide NASA's promotion of the space station well into the 1990s: "It is the next logical step."[1]

These two individuals—one heading NASA and one reporting to the public—set out a framework for expanding the purpose and meaning of human spaceflight. Arguably even more so than for the shuttle, NASA campaigned for a space station, promoting it with a quiver full of ideas, images, and icons designed to build broad support and ultimately win presidential and congressional approval to proceed. The media played their own parts in the campaign—first as explainers, then as critics.

After having won approval in 1984 to develop a space station, NASA nearly lost it repeatedly in the 1990s, in the early 2000s, and near the end of the shuttle program ten years into the new century. As ambitions collided with realities, NASA struggled to maintain a convincing argument for a permanent human presence in space and stayed on the defensive in efforts to convey its message to the public. Meanwhile, influential voices in government and the media assailed the argument as specious. During some twenty years of dueling rhetoric, what NASA called the key to the future, others called a delusion. When NASA framed the space station as an investment, critics rebutted it as an extravagant waste of money. Depending on point of view, human spaceflight was essential or not worth continuing.

Campaigning for "a permanent presence" became a perpetual effort in reframing the concept to accommodate external political pressures and critiques. Under duress, NASA saved its flagship program by scaling back its ambitions and revamping its rhetoric. The *New York Times* became one of the most strident channels of opposition to the space station and the ideas it embodied, challenging it through news reports and editorials. In almost constant crisis over the space station, NASA's persistent campaign to defend the meaning of human spaceflight suggests either a flawed communication strategy or a determined adaptation to opposition.

Most "new ideas" have preludes, and NASA's space station was no exception. In-house, the concept of a space station dated to the origins of the agency, inspired by the visionary thinking of Willy Ley, Wernher von Braun, and others published in the *Collier's* magazine special issues on spaceflight in the early 1950s. Much earlier, in the late 1800s, Russian scientist Konstantin Tsiolkovsky began to think about a space station and then published technical studies on spaceflight and space stations from 1903 into the 1930s. His work became known in the United States in the 1950s and 1960s as foundational thinking about the purpose and design of space stations. Two other European technical thinkers of the early twentieth century, Hermann Oberth of Germany and Hermann Noordung of Austria, conceived of space stations as the basis for interplanetary travel. Von Braun and others knew of, and were influenced by, their work. The idea of a space station was conceivable enough in the 1950s that NASA included a permanent near-earth orbital space station—with a target completion date of 1965–67—as a goal in its first long-range plan, issued a year after the agency's founding. A space station seemed such a logical goal that NASA and members of the aerospace industry convened a symposium in 1960 to air ideas and concepts, and space station planning continued behind the scenes throughout the decade as the manned spaceflight program evolved.[2]

The path of spaceflight in the 1960s led to the moon instead of a space station, and the idea was necessarily set aside under pressure of the space race. However dormant it may have seemed, the concept of a permanent space station never died, and NASA revived it in the late 1960s as one of the primary goals of the post-Apollo program. NASA was not able to win political support for embarking on both a space station and a space transportation system to serve a station, so it postponed the station in favor of developing the space shuttle first. Meanwhile, Apollo hardware and capabilities were repurposed into Skylab, an experimental interim space station that three crews occupied successfully in 1973–74.[3]

NASA issued one of the first salvos in the campaign for a permanent human presence in space as an educational publication, *Space Station: Key to the Future*, in 1970. This booklet presented the rationale for a space station and articulated most of the themes that NASA would use in the 1980s when waging its campaign in earnest. The main theme was practical, productive utilization of space from an orbital facility to be used for ten or more years. It would be a multipurpose, evolutionary, economical research center and a vital link for further exploration of the moon and beyond. A very ambitious hypothetical concept with multiple decks, rotating elements, and a nuclear power reactor, a space station would host research in various fields, sparked by the "critical roles of man," and would open spaceflight to people other than test pilots. The human ability to observe, correlate, learn, respond, adjust, overcome, innovate, improvise, repair, and communicate with insight and skill would be essential for increasing knowledge in a hands-on research environment off the planet. A permanent manned space station would be the key to NASA's future as well as "the" future. Eleven years passed before the real campaign began, but the "big idea" for the next era of human spaceflight had been conceived. During the 1980s NASA would issue a steady stream of advocacy brochures and tantalizing artist concepts to persuade its many constituents of the need for a permanent human (American) presence in space.[4]

Almost from the moment of the space shuttle's first launch in 1981, the highest NASA officials began the effort to win approval for a permanently occupied space station for the 1990s and beyond. One who had close insight into this campaign was its co-architect and next-most-ardent advocate, the NASA deputy administrator from 1981 to 1984, Hans Mark, who collaborated with administrator James Beggs on strategy and tactics. Mark's memoir traces the formulation of their goal and message from the day they were appointed to lead NASA. Public articulation began in their June 1981 confirmation hearings and matured through the process of task group planning, advisory councils, budget deliberations, courtships to win allies, and defenses against opponents in an unrelenting campaign to win the approval of President Ronald Reagan and the Congress. Howard McCurdy's scholarly account of this campaign supplements and enriches Mark's space station story by detailing the behind-the-scenes political maneuvering directed toward the policy decision. Beggs and Mark energized NASA's space station campaign, laying the foundation for efforts that continued after their departures from the agency.[5]

By starting so early to push for "the next logical step" in the future of human spaceflight, NASA faced what could only be characterized as an uphill battle, given that the current step—space shuttle operations—had barely begun. Yet Beggs and Mark looked to the shuttle to play a major role in inspiring public and political support for a space station. As the shuttle made spaceflight routine and demonstrated the potential of human activity in space, they thought it would be relatively easy to urge the political process to the next step. After all, Spacelab, a module carried in the shuttle's payload bay, was essentially a mini–space station, a precursor to the kind of laboratory that would be at the core of a space station. They thought the shuttle and Spacelab missions could be effective sources of public inspiration and political persuasion as to the value of a permanent human presence in space.

Like the space shuttle, the space station imaginary embraced a host of ideas and beliefs. The campaign for its approval exemplified strategies of framing and reframing the meaning of spaceflight to appeal to the right constituents with the right ideas at the right time. While "the next logical step" and "a permanent presence" remained NASA's slogans for years, space station proponents played variations on these themes, accenting certain notes to harmonize with the different interests of their constituents. As in a political or marketing campaign, they strategized, created a narrative around their message, branded this ideology with mottos and logos, and used words and images to convey the ideas and ideals they sought to communicate. In its highest sense, they used rhetoric—the art of persuasion—to mobilize support. As in seeking consensus on the space shuttle, NASA built high expectations that became hard to meet, risking its credibility and fueling opposition.[6]

It is not surprising that an engineering organization headed by an engineering-trained business manager would adopt the conceit of a "next logical step" to frame the next great goal in space. Engineering is a highly methodical discipline of step-by-step creation and problem-solving to design solutions that meet requirements most effectively. Beggs quoted a line from America's rocket engineering pioneer Robert H. Goddard as inspiration for the space station campaign: "Real progress is not a leap in the dark, but a succession of logical steps." Von Braun had envisioned a space station since the 1950s and argued for the logic of pairing a shuttle and station as NASA's post-Apollo space program initiatives. In the sequence of space projects—Mercury, Gemini, Apollo, Skylab—and the long-range plans of 1959 and 1969 that plotted a course of human spaceflight from space station to shuttle (or vice versa) to a moon base and on to Mars, the space station was indeed the long-postponed but next logical step toward a permanent human presence in space.[7]

But NASA's Beggs was not thinking solely as an engineer within the space program context when he revived the space station idea, dormant since the decision to build only the space shuttle. He took a longer, broader view of American culture and world civilization and sought to pin this logical step to beliefs and values beyond the scope of engineering. He saw the permanent human presence on a space station as a step in the evolution of flight, the technological evolution of human civilization, the evolution of humankind as an exploring species, and the evolution of freedom. He remarked that the shuttle and other great creations of man "express not only the cutting edge of our technological skills, but the deepest yearnings within us. . . . They represent a synthesis of technology and humanism, a melding of our highest technical capabilities and of the deepest impulses in our nature." Beggs placed the space station in a cultural context that resonated with the appealing and optimistic idea of progress.[8]

During his five-year tenure at NASA, Beggs tirelessly promoted the space station and elaborated its meaning for humanity. A permanent human presence in space would open opportunities in science, technology, and commerce. It would invigorate, stimulate, and enable activities limited only by the imagination. Human work in space would yield benefits to improve life and solve problems on earth, and it would symbolize peaceful international cooperation. It would set the course for humanity's future into the next generation and the next century. Beggs was less concerned with the architecture of a space station than its human purpose and potential. He always kept the focus on the station as a technological advance that enabled human advances.

The space station campaign began with a wish list of everything an orbital outpost could be—that is, every kind of work that humans could do in space. Some roles on the list originated in von Braun's reasoning in the 1950s and 1960s about the expansion of human activity in space to do research, work on satellites, and stage missions to other destinations. Just as NASA had conceived the space shuttle to satisfy a variety of purposes, so too a multipurpose station would be a versatile resource. An orbital station would serve as a laboratory for doing experiments that take advantage of the microgravity environment; an observatory for unhindered celestial and earth viewing; a processing plant for industries; a technology development test-bed; a satellite servicing center; a spaceport for shuttles and inter-orbit tugs; an assembly depot for large payloads; a staging base for journeys elsewhere in the solar system; a hub for modules from other nations; even a national security platform. Such an ambitious rationale suggested the scope of potential opportunities to be opened by a permanent human presence in space and also suggested that the space station offered something for everyone. Planners came up with more than a hundred credible missions for a space station in its first decade in orbit.[9]

Implicit in such a rationale for a space station was a radically different approach to human spaceflight. Space operations would move from the ground up, into earth orbit. Instead of thinking in discrete missions, or even a series of related missions, as planners did from Mercury through the shuttle, space station planners essentially foresaw the whole space enterprise increasingly based in orbit. A space station would be more than an extended foray in living and working off the planet; it would be an incipient space colony. Set in the frontier paradigm, the space station would be the early settlement that leads to expansion and commerce and becomes ever more autonomous. The station would become the leading edge of humanity's movement into space.

In theory, framing the space station as a multipurpose, opportunity-rich facility should attract the broadest range of interest as a basis for building a coalition of supporters. This approach would enable NASA to identify potential users and their technical requirements before presenting a design concept that would surely spark debate. This strategy intended to avoid crippling the project too soon with political bargaining over its size and capability. Such a strategy ran an opposite risk, however, that became evident several years later: the more diverse the functions of the space station became, the more complex and expensive its design became, until the concept began to collapse under competing pressures and compromises.[10]

Beyond its capabilities, another element entered the rationale for a space station: leadership. By 1982, the Soviet Union had sent multiple crews to its sixth Salyut space station, opened Salyut 7 for occupancy, and announced plans for a much larger station in the 1980s (which became *Mir,* established in orbit in 1986). NASA and the media could not ignore that the Soviets had built a long lead in human spaceflight with a decade of experience in orbital stations and record-setting stays in space while the United States had been busy developing the shuttle. Advocates for a U.S. orbital station began to echo space-race-era appeals to national prestige and the call of freedom. "The next logical step" also became a step toward reasserting leadership in space: America should not be left behind while the Soviets reaped the many benefits of a permanent human presence in space. Beggs remarked, "A space station must be built if we are to maintain the position of leadership in space. . . . The Space Shuttle still gives us the edge . . . but alone, the Shuttle will not enable the United States to realize the full potential of space. Only a space station . . . can do that."[11]

Hans Mark recalled the space station campaign as "a lengthy process of inspiration, technical planning, and political persuasion." Inspiration was perhaps the most challenging part of framing the meaning of human spaceflight because inspiration arises from intangible values and emotions, whereas technical planning is largely factual. "Selling" and "pitching" are often shaded as pejorative terms for persuasion, but strategic messaging is imperative in any effort to win support. Shortly after assuming their positions at NASA, Beggs and Mark established parallel advisory efforts on both the strategy for winning approval and the technical requirements for a space station.[12]

The rationale for human spaceflight would be a key issue again for NASA, as it had been in winning approval for the space shuttle years earlier. Skeptics and opponents immediately raised questions about the need for a permanent *human* presence in space when machines can do much of what people want to do there. The scientific community argued that much research in space could be automated, and the defense establishment had no national security requirements for humans in space. The "manned vs. unmanned" debate over costs and competing priorities inevitably arose. Yet NASA's advisory committee unequivocally recommended that any space station should have people aboard from the start, because anything less would not be bold enough to garner political support. Nor would anything less realize the full potential of space operations made possible by the shuttle and inherent in a space station. The optimistic themes of leadership, progress, and the evolution of civilization assumed a human presence in space.

NASA chartered a Space Station Task Force led by John D. Hodge to develop the concept of a space station—what it could be, what it could do, what it would cost—to support NASA's efforts to win approval. The task force focused on defining the fundamental functions and user requirements of a space station and did solid technical work for two years (1982–84) to produce the six-volume *Space Station Program Description Document* that would help make the concept a reality. The task force also served as the campaign headquarters or "war room," analyzing issues, scoping opposition, strategizing, and preparing persuasive stump speeches, slogans, and literature. Its tasks were both technical and rhetorical, and its products reveal the work of crafting the elements of persuasion. This team originated much of the space station ideology and iconography.[13]

At first the task force used Beggs's paper, "A Space Station for America," as the master script for the space station rationale, updating and reissuing it as a printed pamphlet, always ending with the phrase "the next logical step." They fine-tuned its message with subtle changes in each version, adding or omitting points and editing nuances in an exacting process of message framing. All versions made the point that space was competitive and U.S. leadership there no longer went unchallenged; America *needed* a space station to remain a leader. As exercises in communications, they reveal the careful crafting of language to frame the meaning of a human presence in space.[14]

Two of the fifty or so members of this task force played key roles in framing the messages of the space station sales pitch in other formats. Robert F. Freitag, Hodge's deputy director, was an engineer, planner, and manager of new programs in the Office of Manned Space Flight. He also was a writer and a firm believer in the space station. Political scientist Terence T. Finn served as director of policy and plans on the task force. He had ten years' experience on the staff of congressional committees and a passion for the workings of government. They worked closely with Hodge to identify the themes and write the scripts to tell the space station story outside—and even inside—the halls of NASA. The various publications explaining and arguing for a space station issued by NASA in the 1980s were largely their handiwork.[15]

Finn especially was known for his "Think Pieces" and memos strategizing how to further the cause of the space station. A tireless activist in the hands-on work of framing the meaning of a space station, he mapped out the routes to NASA's constituencies and prepared lists of who should be doing what and talking to whom to get the word out. He kept tallies of what NASA was doing well and not so well and suggested actions to keep the effort on track. He developed a "polished two-pager" for Beggs to leave with his contacts in the White House and Congress. A stickler for staying on message, Finn insisted on a set of themes and bullet points because everyone should say (or not say) the same things, and he worried that the agency might not be disciplined enough to wage a successful campaign.[16]

In addition to publications, the Space Station Task Force assembled a collection of visuals to tell the space station story. NASA managers could select and shuffle these charts and images to tailor talks to audiences as varied as the NASA Advisory Council, the NASA center directors, other federal agencies, congressional committee staffers, aerospace companies, professional societies, and civic groups. During the intense campaigning of 1982–83, Hodge, Freitag, Finn, and other speakers made about two hundred appearances to advocate for the space station. Finn said that they were "trying to build a constituency for Space Station" and "broaden the base of understanding" that it "was indeed the next logical step to take in space."[17]

Encountering early opposition from key administration officials, Beggs decided to try something different from briefings. He focused on gaining the president's attention directly by engaging him in the experience of human spaceflight. Getting him out of the White House and close to space operations might prime him for a favorable space station decision. Beggs extended, and President Reagan accepted, three invitations to join him during missions in progress; they visited NASA's Mission Control Center during the second shuttle mission, attended the landing of the fourth and final test flight mission, and held a conference call with the international crew of the first Spacelab mission. Each event exposed the president to the heartbeat of human spaceflight in experiences that he visibly enjoyed.

Of the three occasions, the *Columbia* landing was a crucial step toward NASA's goal of space station approval. The scene could not have been better set in a Hollywood film—July 4, 1982: a patriotic holiday crowd along the shuttle landing strip in the southern California high desert, a stage trimmed in bunting, and opportunities for the president to greet the returning crew as they stepped out of the orbiter and then give a speech with them on stage. NASA produced a personal experience replete with symbols—a visual and experiential rhetoric—that appealed to Reagan's values more persuasively than a briefing. The president was the star attraction in an event that brought together patriotism, spaceflight technology, astronaut heroes, and an enthusiastic public. To end the ceremony, Reagan declared the space shuttle operational and watched *Challenger* depart for delivery to Florida as the second orbiter in the fleet. It was a moment bright with promise for the shuttle era of routine spaceflight and the nation's future in space.

NASA had worked hard for the president to announce a commitment to build a space station in this speech, and Mark had in fact written suggested remarks that included such direction to NASA. The president's science adviser and the budget director deemed an announcement premature and forbade any direct reference to a space station. Accounts of the speechwriting process indicate a struggle over exact wording. NASA insisted on at least an allusion to "permanent presence," and White House staff adamantly opposed the words "space station." The compromise became a statement in the speech, and in the national space policy document released on the same day, about establishing "a more permanent presence" in space.[18]

Although not the firm commitment NASA desired, the phrase "a more permanent presence" was encouraging. A comparison of Mark's draft with the actual speech shows that the White House version echoed NASA's human spaceflight rationale. After an opening image of gallant Yankee clipper ships returning home to port, Reagan spoke in his preferred rhetoric of the frontier. Of the space program and shuttle, he said, "The pioneer spirit still flourishes in America," and he spoke of challenges, freedom, exploration, discoveries, commercial ventures, innovations, knowledge, benefits and betterment of life on earth, science, private enterprise, national security, international cooperation, peace—all the themes that NASA embraced within the "next logical step" ideology. Although President Reagan was not yet ready to make a space station decision, the speech made it clear that he understood the meaning of human spaceflight in essentially the same vocabulary that NASA did. According to news reports, Reagan was awed by this encounter with the excitement of spaceflight. The events and speech of that July 4 were crucial in shaping both the president's personal awareness of spaceflight and NASA's continuing campaign for a space station.[19]

Beggs cleverly used the new icon for human spaceflight—the space shuttle—to set the stage for the space station decision. The impact of Reagan's presence at the landing gave immediacy to the idea of a "next logical step." It was not a quantum leap of faith to understand that this vehicle needed a place to shuttle to and from. Since the 1960s, planners had paired the shuttle and space station, and now that rationale was visibly evident. Likewise, as the tally of successful shuttle missions began to grow, so too did the wealth of images depicting what people were doing in space. The icon and images made spaceflight seem normal and routine enough that living and working on a space station was a reasonable next step. The campaign for a space station benefited from imagery and reality, whereas proponents of the space shuttle a decade earlier had to depend on words and artists' renderings. NASA now had the ability to augment words with a compelling visual rhetoric to shape the meaning of human spaceflight.

One gauge of the effectiveness of intended meaning is its appearance and repetition or critique in the media. As in its coverage of the dawning space shuttle era, the *New York Times* actively reported and commented on the proposed space station. The point of view was not uniform, however. John Noble Wilford wrote with guarded optimism about the potential of a permanent human presence in space, because "we are a spacefaring people who, in all likelihood, are going to make ourselves at home in space someday." He acknowledged that

NASA needed to move beyond "dreams and schemes" to a detailed long-range plan and reduction of the cost of spaceflight, but he did not initially share the skepticism of some of his colleagues or members of the newspaper's editorial board. He admired the "startling virtuosity" of the U.S. space program, a quality that argued for clearly defined long-range goals for human spaceflight. He "bought" the leadership argument for the next logical step in space.[20]

The *New York Times* editorial board made little comment until late 1983, when it was clear that a decision was imminent. It sided with President Reagan's science adviser George Keyworth in charging NASA with a "lack of imagination" and noted that neither the National Academy of Sciences nor the Pentagon saw a need for a space station. In the editorialists' view, "a big space station would be just an orbiting white elephant unless its purpose were carefully defined." Even ten years later, the *Washington Post* ran an editorial cartoon variant of the same meme—the space station as a white elephant tethered to a bag of money above a critical article about the "Megabuck Space Station." Clearly, NASA still had prolonged work to do to defuse that kind of critique.[21]

NASA's campaign to win approval to build a space station peaked in 1983 with intense efforts to influence key members of the Reagan administration through a formal policy development process. The Senior Interagency Group for Space comprising representatives of NASA; the Departments of Defense, State, and Commerce; the Central Intelligence Agency; and the Arms Control and Disarmament Agency carried out a study to assess NASA's space station proposal, and any alternatives to it, as a basis for the president's decision. Of these, only Commerce was favorably disposed toward a space station. NASA's space station advocacy team tried to "push it through," according to Hans Mark, but did not succeed in converting opponents. There was too much opposition on fiscal issues and a sense that a space station was premature and unnecessary, and too little support from agencies having no uses for a manned space station. The study group could not reach a consensus or recommendation for the president's consideration.[22]

Framing the space station effectively meant getting the arguments right and also finding the right channel to capture the president's attention. As the formal policy process foundered, NASA's administrator sought to open other paths through the White House. Near the end of 1983, Beggs managed to present the case for an American space station to a policy advisory council on commerce and trade in their meeting with the president. There he stressed the economic points that had not persuaded the interagency group, showing that commercial activity in space served the national interest. At the time, the Reagan administration was developing a commercial space policy to stimulate greater private enterprise. Taking a cue from this initiative, NASA enhanced its argument for the use of a space station for commercial activities and economic benefits.

Beggs's presentation notes for that meeting revealed how meticulously framed the space station argument had become for this critical moment, and Finn later explained how they "very, very carefully constructed" the briefing to appeal to "a traditional yearning to be great and to be the best." The themes of "A Space Station for America" were rearranged and rebalanced for greater appeal to a business-minded audience and to President Reagan's interests. Beggs led with the importance of leadership, remarked on the successes of the space shuttle, segued to the "next logical step" argument, focused on commercial private-sector opportunities, extolled space as a place to do business, listed types of work that could be done on a space station, argued that people are essential in space, raised the specter of uncertainty about Soviet intentions in space, picked apart the options that the interagency group had considered, mentioned decisions by Presidents Kennedy and Nixon for U.S. preeminence in space, and returned to the leadership theme. He ended dramatically: "The time to start a space station is now. . . . The stakes are enormous: leadership in space for the next 25 years." Within a few days, Beggs received word from the White House that President Reagan had decided to commit the nation to building a space station.[23]

President Reagan included the announcement in his 1984 State of the Union Address, an event that culminated NASA's two-year campaign. As polished as NASA's own message had been, the president heightened its impact by framing the space station as an emblem of America's greatness. Like inaugural addresses, the annual State of the Union Address to Congress is carefully crafted to articulate the vision and values that the president claims for his legacy. More than most presidential speeches, this address is expected to inspire, to soar. As always, Reagan situated spaceflight in the context of pioneering a great frontier. Then came the memorable line—"America has always been greatest when we dared to be great. We can reach for greatness again"—that served as a prelude to his direction to NASA to develop a permanently manned space station. The themes of commerce, private enterprise, new knowledge, international cooperation, peace, prosperity, and freedom ran through the few paragraphs dedicated to this new goal in space. President Reagan had gleaned from NASA the essential idea of leadership and enhanced it to inspire the nation toward greatness again in space.[24]

Press coverage of the State of the Union Address paid due attention to the bold space station announcement but also noted that congressional support must be won. One reporter noted wryly that a space station had been approved that apparently no one but NASA wanted. Why? Because the president liked it.

The next day's *New York Times* editorial called the space station "an expensive yawn in space" and dismissed "its striking lack of imagination or technological challenge." That newspaper's consistent editorial theme became NASA's obsession with human spaceflight and the contrast in efficiency between manned and unmanned missions. In many editorials and opinion pieces throughout the year, it claimed that unmanned programs offered "far greater opportunities for stirring the public's imagination" or could do anything humans could do more reliably and economically. Despite NASA's calculated persuasive efforts, the *Times* argued that the space station idea was "old hat," "the ultimate junket," "the wrong stuff," and a "big ticket, make-work program to keep the space agency [and the shuttle] busy." Among its scornful pronouncements, this one pointedly dismissed Reagan's endorsement: "The problem with NASA and a manned space station is that it does not dare . . . it has barely as much challenge as putting a new base in Antarctica, as much romance as turning astronauts into teamsters." In the face of such negative rhetoric, NASA's campaign for wider support was not over.[25]

With a green light from the president, NASA proceeded to open the space station program office, begin the contracting process, and go into business to develop a space station. But in gaining White House backing, the campaign for a space station had passed only the highest hurdle. Congress presented other challenges, and Beggs and the space station program team continued to make the case for a permanent human presence in space. They parlayed the president's charter into their congressional testimony, a blitz of articles in the aerospace press, and an updated master script. This one, also issued as a NASA publication in Beggs's name, refreshed the main idea: *Space Station: The Next Logical Step*. For the first time, it presented the case for a space station with colorful illustrations. The brochure stressed the symbolic value of a space station without arguing for a particular design and accented themes that would resonate with aerospace industry and political leaders—revitalized American technology, competitiveness, and private-sector opportunities in space. Beggs appealed to national pride in this credo: "Starting in the early 1990s, I believe there will always be Americans living and working in space."[26]

Just two months after Reagan's 1984 State of the Union Address, Beggs, in testimony before the House Appropriations Committee, identified eight functions that a space station would serve. These differed somewhat from the ones listed in his 1982 position paper: he dropped any mention of the Department of Defense and national security and separated some previously clustered functions for greater visibility—a laboratory, an observatory, a transportation node for missions to other orbits, a servicing facility, an assembly facility for large structures, a commercial manufacturing facility, a storage depot, and a staging base for more ambitious future missions. His testimony represented an expansive vision of human activity in space that embraced an ambitious variety of scientific, technical, and commercial opportunities. Curiously, it failed to mention the obvious: the station would also be a home for the people who would do all this work.[27]

As NASA hired contractors for a three-year space station definition effort, they kept the public's focus on general principles, not on specific design issues. To avoid early arguments over design, Beggs decided that NASA would spread a broad message that user requirements would shape the design (making it customer-friendly), and the station would grow in size and capability (evolve) as needed. Instead of focusing on architecture, NASA emphasized the role of the space station to provide people with "an opportunity to extend their wisdom and skills to new frontiers and horizons, to stimulate their imagination, to generate new ideas, to nurture their creativity, and to benefit intellectually." This vision of the human potential of living and working in space had a radical goal: to become liberated from thinking and operating as "one-g machines" bound to earth—to begin living and working in space (zero-g) full time.[28]

The "next logical step" theme headlined increasingly elaborate NASA publications during the period of space station definition and preliminary design studies from 1985 through 1987. NASA's announcement of a dual-keel design in 1986 (described later in this chapter) sparked a variety of efforts to publicize progress toward an actual space station. The agency issued at least two major promotional-educational pieces in full color, and the modest black-and-white NASA Facts and Information Summaries series also joined the chorus. All promoted a permanently manned space station as the next logical step and continued to tout its multipurpose potential. These publications also began to depict the features that would be provided for crews in a habitat module and to use photos from shuttle missions to show how astronauts might assemble and service a space station. A new theme based on the logical step appeared—the space station as a "stepping stone to the future" that would help make possible human expeditions well beyond earth orbit. As in earlier pieces, the related themes of U.S. leadership in space, competitiveness, and the symbolic value of such a complex endeavor threaded through the texts.[29]

Of these 1986-vintage publications, the most substantial was a fifty-page color illustrated expanded edition of *Space Station: The Next Logical Step* with images of people living and working in space (plate 10). As an element of NASA's campaign strategy, this booklet deftly communicated NASA's updated main themes.

It emphasized a permanent presence in space for research and other work, the potential rewards for science and commerce, the satisfaction of national priorities, and the benefits of international cooperation. The space station would be a symbol of American resolve, innovation, and national pride. Relevant quotations of historical and political figures, including President Reagan and a member of Congress, accented the pages. The booklet also addressed two value-laden matters directly: why a space station now, and what the space station means. The logic of "our step by step climb into space" responded to the "Why now?" question, and the meaning of a space station was to open a "gateway to the future." Like the stepping-stone trope, a gateway set up the expectation of another next logical step in America's progress and brought the space station into a larger scenario, even before it was designed or built. It would "turn humankind from transients to residents of space."[30]

The year 1986 became a pivot point for thinking about human spaceflight for two reasons: the shock of the *Challenger* shuttle launch tragedy and the release of the space station baseline configuration. One challenged the basic precept of routine access to space and prompted public awareness of the risks and roles of people in space. The other established a realizable concept for a permanent human presence in space. NASA adjusted its view of human spaceflight in response to both events.

The *Challenger* tragedy provoked reconsideration of the proper role of humans in space that dogged space station deliberations for years. When were people truly necessary? Was delivering or repairing satellites and doing similar work worth the risk and cost of human spaceflight? Should spaceflight be restricted to missions that only humans could perform, to accomplish complex tasks for which there were no alternative means? These questions related to the nation's broad goals in space and a desire for greater clarity in the purpose of the shuttle, space station, and human presence. Calling the space station "an infrastructure in search of a mission," the *New York Times* dismissed the rationales for human presence as unconvincing and continued to fault NASA for a failure of vision.[31]

The 1986 version of the *Space Station: The Next Logical Step* publication appeared shortly after the loss of *Challenger.* It drew some content from prior versions, but it also sought to address any doubters about the post-*Challenger* future with assurances that the space station "continues to be" the next logical step toward exploration and leadership. The opening words stated emphatically that the United States and the president remained committed to develop a permanently manned space station, with the final message: "Now more than ever, the Space Station is important to our future and we must move forward as planned."[32]

With release of the space station baseline configuration, attention shifted from grand themes to the more pragmatic matter of architecture. By 1984 NASA's space station task group studies had resulted in a preliminary design concept, the "power tower," that clustered laboratory and observatory elements on a long keel topped by solar panels. The contractor studies that began in 1984 yielded, in 1986, a revised concept, the "dual-keel" space station, which allowed for more attached payload platforms and a more efficient arrangement of laboratory and living modules. This became the baseline configuration for further design refinements and for the continuing discussion of human activity in space.

Subjecting the dual-keel configuration to more technical scrutiny and cost analysis became a series of paring-down exercises. Emphasis shifted to architecture more than the human role in space. What configuration of modules, platforms, solar arrays, and other features would be most conducive to utilizing space while staying within funding constraints? Inevitably, over the next few years design followed redesign, and some of the human spaceflight functions were lopped off. The habitation module—the human crew's living quarters—was among the first elements to be scaled back during these redesign (downsizing) efforts. This decision essentially meant that the needs of the humans present in space were dispensable compared with the political and budgetary requirements on earth. To meet the demands of the political process, space station advocates had to sacrifice some of the station's versatility and thus the range of human activity in space.[33]

The agency released a revised baseline station concept in 1988, the same year that international partnership agreements were executed. By then, the station was becoming home to a highly specialized human presence that would carry out certain kinds of scientific research and assemble, operate, and maintain the facility. It would no longer try to accommodate every purpose. Even so, the station was still faulted as grandiose, extravagant, lacking clear goals, and deserving postponement or sudden death. An exasperated *New York Times* editorial board futilely demanded that Congress "cancel this celestial circus."[34]

As the space station concept matured, proponents urged a less generic name than space station, arguing that it needed an identity as something distinctive and real that would capture and hold attention, even inspire support. Earlier, a reader had written to an aerospace journal that "we might make more progress if we would stop calling it a 'space station' and start calling it what it really would be—a 'research center.'" He suggested dropping the words "space station" altogether. Critics might object to NASA building a space station, he noted, "but they would have to agree that NASA is chartered to build 'research centers.'" He further suggested the name Neil Armstrong Research Center to honor America's space hero, just as the main NASA centers on earth were named for great men. Anonymity was bland, but a respected name might help motivate support for a space station.[35]

From time to time NASA had received inquiries about and suggestions for naming the space station. In 1984, a magazine solicited readers' ideas and the publisher reported an enthusiastic response, forwarding a list of several names to Administrator Beggs for consideration. *Pegasus* was the readers' preferred choice, but *Genesis, Capricorn, America,* and some others had a following. In 1986, there was some public sentiment for naming the space station to honor the last *Challenger* crew—perhaps Challenger Memorial Base—or holding a contest for students to name the station, but NASA was not yet ready. A Congressman opined that "space station" sounded "like a floating gas pump," making it much easier to slash funds from the project, whereas a more appealing name could win converts. For some time, though, descriptive terms—first "power tower" and then "dual keel" or simply "space station"—sufficed for NASA.[36]

By 1988, however, with the initial contractor studies completed, a basic concept had emerged, and it was time to begin designing the space station in detail. More important, it was time to move the space station budget request through Congress. Word went out from NASA Headquarters to all the NASA centers: "The Space Station needs a name." Now was the time for an identity. Evidently, supporters in Congress felt that a "real name" would help the program through the appropriations process, so NASA swung into gear to find a name that sounded good, was appropriate, and "wouldn't offend anyone." The process began with a managerial meeting at NASA Headquarters in April to develop a list of possible names to be sent out to the centers for comments and more suggestions. Then a committee at Headquarters would review the feedback and recommend several finalists for the administrator's decision. Two early favorites were *Freedom* and *Pilgrim,* both resonant with American ideals.[37]

NASA's offices of external affairs, public affairs, space flight, and space station organized the process and criteria for generating a selection of suitable names. The solicitation netted more than five hundred submissions of names drawn primarily from Greek and Roman mythology, astronomy, and American history. The most frequently suggested name by far was *Genesis;* multiple suggestions also arrived for *Frontier, Oasis, Skybase,* and *Starbase.* The space station naming committee at NASA Headquarters winnowed the list to about fifty names and then further winnowed it to derive the top candidates. By the end of May 1988, the committee was ready to present the NASA administrator four names for consideration. *Pegasus:* like the mythical winged horse among the stars, the winged space station would be the steed for human exploration and exploitation of space. *Odyssey:* from the space station, the ultimate human adventure and journey to explore the frontiers of space would begin. *Freedom:* the space station would represent a basic human desire common to all people, a political value central to the space station partners, and the opportunity to do research in space without the confines of gravity. *Friendship:* the spirit of international partnership to design, develop, operate, and use the space station. *Orion* and *Olympia* followed as alternates. James Beggs had left the agency in 1985, so the name choice fell to his successor, James Fletcher, serving his second term as NASA administrator.

Further vetting of these candidate names with the international partners revealed nuances to be taken into account. The Europeans preferred *Pegasus* and *Friendship.* The Canadians preferred *Odyssey,* but the Japanese were concerned that it connoted "wandering." *Pegasus* might be confused with a new U.S. commercial launch vehicle of the same name, and *Friendship* was a trademark call signal of the Russian airline Aeroflot. *Freedom* might suggest a response to the Soviet presence in space. As a result of these nuances, the committee evidently revised the selection for final consideration, because media reports indicated that the finalists were *Freedom, Aurora,* and *Orion.*[38]

During June, NASA sorted out the preferred names and sent a recommendation to the White House. That same month, the space station international partners completed negotiations on the intergovernmental agreements for participation in the largest cooperative scientific and technological project ever undertaken in space. This milestone made an auspicious time to announce a name for the space station. The White House staff memorandum seeking President Reagan's decision suggested that naming the space station then would reaffirm his support at a crucial time of funding difficulties in Congress and would heighten public awareness and appreciation for the project. The memo recommended selection of the name recommended by NASA and the international partners, a name that "conveys the appropriate image for the West's space station, and complements nicely the Soviet name for their space station, *Mir,* which translates as 'Peace.'" On July 18, 1988, President Reagan announced the name of the international space station: *Freedom.*[39]

The rationale for the name *Freedom,* scripted within NASA, appeared in the White House press release. "The yearning for freedom is a basic human emotion, and freedom of the individual is a value shared by all the nations that will work together to build and use the Space Station." Offering freedom from the confines of earth's gravity, the space station would enable research, commercial activity, and human exploration in space, its name thus conforming to NASA's main goals for a permanent human presence in space. The name *Freedom* resonated with Reagan's words in his 1984 State of the Union Address, when he envisioned the space station as a way to strengthen peace and expand freedom. In the public psyche, this name had both political and patriotic strength. NASA could move its space station campaign forward with the *Freedom* banner.[40]

NASA soon published style guidelines for use of the space station *Freedom* name and developed a graphic logo (fig. 5.1). Both were standard marketing tools for establishing product or corporate awareness through a recognizable visual identity. Like the naming process, this attention to graphic identity and style guidelines indicated the scope of NASA's deliberate efforts to frame and brand its space station message. Creation of a graphic icon to appear on stationery, publications, briefing charts, decals, lapel pins, and other materials gave the agency additional ways to make an impression and increase visibility of the space station program.[41]

With an agreed-upon basic architecture and name for its space station, and just enough funding to start planning the details, NASA did not desist from campaign mode. In fact, the agency continued to release a stream of materials to sustain the momentum of a permanent human presence in space. The master theme, however, began to shift from "next logical step" to other appeals after the president and Congress approved the space station development. The future became the dominant new theme, with the space station often characterized in architectural terms as a "gateway," "keystone," or "stepping stone" to the future or a "foothold on" the future. The corollary to this theme was leadership. NASA publications issued from 1987 through 1992 typically presented the argument for a permanent presence in space as the means of enabling whatever future goals in space the nation might choose, while also ensuring leadership in scientific research and technology.[42]

Terence Finn assembled for the space station office a set of themes for talks and publications. Leadership in space topped the list. Permanence was another

Fig. 5.1. This logo conveyed the essence of space station *Freedom*'s design simply, without symbolism other than its auspicious name. NASA (89-34285).

key point: "No longer will we visit space. We will be there, living and working, all the time." The station's evolutionary design also was a selling point related to permanence; the station could change and grow as technology and human interests changed and as funding continued. He advised stressing the user-friendliness of a space station, implying the breadth of access to space for practical, beneficial purposes. The president had insisted on international participation for peaceful activity in space, another important point. Even with the program started, the annual budget process meant that the campaign would continue for years, and the space station advocates must keep their message focused yet refreshed. Finn made it his mission to be steward of the ideas and chief communicator to keep everyone on message.[43]

The aerospace giants who vied for space station contracts also joined the effort to persuade Congress and the public of the merits of a space station for a permanent presence in space. Boeing, Lockheed, Martin Marietta, McDonnell Douglas, and Rockwell published colorful brochures and advertisements to spread the message that America needed space station *Freedom.* These appealed to business values of innovation and competitiveness, quality of life improvements and spinoff technologies in various commercial sectors, return on investment, and contributions to a healthy economy. Boeing ran a particularly emotional series of print ads featuring a child with an incurable disease who might be helped by research on the space station or a young adult on the cusp of a career under the slogan "Space research is this generation's call to greatness." Industry promoted the argument that a permanent presence in space (a space station) was an investment in national strength, global leadership, and the nation's future. And, of course, it would be a catalyst for jobs and prosperity.[44]

Two NASA publications from the space station *Freedom* period were especially noteworthy for their heft and sophistication. Whereas the earlier *Space Station for America* and all but the 1986 *Next Logical Step* booklets were fairly simple—a few pages of type with a pictorial cover and a few internal images—the ones circulated for the next five years were more comprehensive, more colorful, better illustrated, and as stylish as magazines. The Office of Space Station hired a space journalist to write a forty-four-page volume titled *Space Station Freedom: A Foothold on the Future* that addressed all the key themes. Full of information but reader-friendly, the booklet described the elements and functions of the space station, but more importantly explored the human presence and its meaning in a new era of living in space. The volume depicted in words and images what this stage of "our permanent settlement" in low earth orbit would look like. The earlier publications were about selling the idea of a space station; this one was about turning the idea into reality and creating a visual culture around it. The booklet featured large illustrations by notable space artists Robert McCall and Vincent Di Fate and many photos. This material essentially invited the reader into the space station, explained the accommodations for living and doing research, and looked to the future when dreams of traveling into the solar system might be fulfilled from the space station. The author (and those who commissioned him) presented space station *Freedom* as a symbol of the future.[45]

The second prominently glossy publication, issued in 1992, dubbed space station *Freedom* the "Gateway to the Future." It opened with claims that "Great Nations Dare to Explore" and "Space Station *Freedom* will open a new era of exploration." It depicted the station as the gateway to space, to scientific research, discovery, utilization, and benefits. Most of the pages featured human activity in space, characterized as exploration. This shift in meaning upped the ante on the meaning of spaceflight at a time when negative opinion held sway in the political realm and the space station barely escaped cancellation. Editorial opinion was similarly unfavorable, with the *New York Times* insisting that NASA should have a worthier goal than staying in earth orbit and that robots made better explorers than people. The "Gateway" publication somewhat disingenuously countered that argument by recharacterizing the role of people on the space station as explorers, albeit as seekers of knowledge (confined to earth's neighborhood) rather than venturers to distant places. This new layer of meaning suggested resilience in framing the space station idea, but the change in emphasis may have been more defensive than proactive, adopted to fend off mounting assaults by critics.[46]

Despite the president's endorsement in three State of the Union Addresses, the *Freedom* years were fraught with challenges, and NASA's upbeat case for a permanent presence in space belied the growing opposition. The period 1984–90 saw cycles of spiraling growth in the space station's size, technical specifications, and projected cost, from $8 billion to $16 billion to $38 billion, and subsequent downsizing. Constrained funding to curb this growth, plus pressure from the administration and Congress to reduce the station's scope, kept NASA in an ongoing campaign to protect the station from postponement or cancellation and also to expand support for a permanent presence in space. A prescient 1988 editorial cartoon captured the theme of the next few years; it showed two cups beside a space station blueprint on a drawing board—one labeled NASA, filled with sharpened pencils, and the other labeled Congress, filled with erasers. With each redesign to trim cost, the size and capability of the station shrank and the human potential there diminished. This paradox made it increasingly difficult to defend the importance of a human presence in space and kept NASA in a mode of transforming meaning and message to maintain, if not broaden, the appeal of a manned space station.[47]

In 1989 the agency put together a simple black-and-white six-panel folder that opened into a poster. It could be produced in quantity inexpensively and left all over town. *The Case for Space Station Freedom: A Statement of Purpose* was a blatant effort to shape the meaning of human spaceflight. The main ideas leapt from the headings: "The Next Logical Step," "The Waypoints of Exploration," and "Why Now?" This little publication appealed to both the past ("An orbital station is the natural extension of America's spacefaring enterprise") and the future ("It is a reflection of the kind of legacy the present generation plans to leave"). It urged that "Leadership abandoned is leadership lost" and all but stirred guilt at the thought of delaying *Freedom.*[48]

To keep the space station program afloat in a storm of criticism and pressure to abandon it, NASA had the unenviable task of trimming, even hacking up, its highest priority in order to save it. With thinning political support and tight funding limits, NASA redesigned the station in 1990, 1991, and 1992 to try to fit within constraints set by its congressional oversight committees. Redesigns resulted in a progressively smaller, less capable, and less costly *Freedom* concept, but criticism and controversy continued unabated. The challenges of framing and communicating the space station message in an environment of crisis were enormous.

Meanwhile, unrelenting in its opposition, the *New York Times* ridiculed the station and reframed the issue of human spaceflight in a volley of contrary rhetoric. Calling the space station an "old-hat" concept and accusing NASA of having lost its sense of direction and imagination, editorials argued for intelligent and effective space exploration by unmanned means. The paper expressed a strong bias toward keeping human bodies on earth and letting only human intelligence into space instead of engaging in "the circus of manned space flight." Rather than "musty plans and cramped vision"—a "black hole in space"—the nation deserved a space program that "stirs people's sense of adventure." The space station was nothing but a grandiose contrivance, a "space palace," a "luxury sky hotel," a "castle in the air." It should be postponed or canceled because there was no clear goal for it; "instead of reaching to stretch man's grasp, NASA's engineers plan more plumbing." The *Times* argued that it was time to acknowledge that "the space station makes no sense" and cancel it.[49]

An example of this hand-to-hand war of words was the term "stepping stone." In a space station "Information Summary," public reference material routinely available in NASA press offices and press conferences, one of the sections, titled "Space Station Freedom—Stepping Stone to the Future," stated that *Freedom* was essential for the advancement of human exploration of space. An editorial in the *New York Times* mocked the idea, arguing that the space station wasn't a stepping stone to anywhere and then methodically dismissing most of the allegedly valuable uses of the space station.[50]

To counter this barrage of criticism, NASA issued an updated edition of its *Space Station Freedom Media Handbook.* Here, in about a hundred pages of information introduced by a statement of purpose, the agency made its case. In a quick history of space station concepts and recent changes in design, a new message appeared: that space stations exist for one basic purpose—*to enable exploration*—and that *Freedom* was "first and foremost a means to that end." It would be "an accessible waypoint along the outbound trail" and a "modern equivalent to the river junction" for America's spacefaring enterprise. The rhetoric then situated *Freedom* as the embodiment of ideas from the founding fathers of rocketry, NASA's charter, the Age of Discovery, and the westward movement of the United States. Previous priorities were reversed as research in space became secondary to advancing human exploration. The space station's primary mission would be "a permanent outpost where we will learn to live and work productively in space . . . before the nation can embark on achievable, long-range human exploration goals." It would *also* serve as a laboratory and engineering testbed for the benefit of humanity. The media guide was an unabashed effort at persuasion, claiming that it was time to ride the momentum of thirty years of spaceflight experience into the future, to invest and lead and create the present generation's legacy.[51]

The future of human spaceflight became a tug of war between NASA and Congress over the architecture, cost, and funding of space station *Freedom.* At least twenty-two times, votes in the House and Senate during the early 1990s threatened cancellation or curbed its budget, forcing redesigns to scale the station back from the size of the U.S. Capitol to something more affordable. Editorial cartoonists captured the essence of the redesign saga, depicting the shrinking space station variously as a unit in an orbital trailer park, an airstream trailer, a station wagon, an orbital phone booth, and a desktop model—all the product of budget cuts (fig. 5.2). A cartoon perfectly expressed the discrepancy between the way NASA framed space station *Freedom* and its framing by opponents: in one panel (as seen by supporters) as a key to the future, in the other (as seen by critics) as a pie in the sky. NASA was caught in an increasingly desperate cycle of trimming away at its vision for a permanent presence in space to satisfy Congress and placate critics, yet the more it lopped elements off the station, the more circumscribed its rationale became.[52]

Fig. 5.2. This cartoon satirized the endless redesigns and reductions NASA faced before finally gaining approval to start construction. Cartoon by Dana Summers published in the *Orlando Sentinel*, July 8, 1993. Used by Permission. © 2016 Tribune Content Agency, LLC.

The space station dilemma came to a head in 1993 in a "do-or-die" crisis. A new president, Bill Clinton, ordered NASA to downsize the space station again and reconfigure it to integrate Russian elements and the Russian Space Agency as a full partner in the program. Conceived as an international effort and an instrument of foreign policy since President Reagan's direction, in part to rival the Russian presence in space, the space station now had to serve as a symbol of unity in the post–Cold War world. More important, per the Clinton administration, the constant redesigning and cost escalation had to stop. If there were to be a space station at all, the basics had to be settled in a viable, affordable plan. It would need a fresh name, too, to shed its fraught history.[53]

Relieved that the station would not be terminated, NASA administrator Daniel Goldin ebulliently called this "an incredible opportunity" to bring together NASA's most creative minds to reconceive the space station, even to start from scratch again. This space station redesign team would literally and metaphorically reframe the architecture and purpose of a human presence in space. Goldin may have put the best possible spin on the situation or understated its gravity to boost morale after months of anxiety, but the media read things much differently when the White House made the redesign instructions public. NASA had to develop three options for a reduced-cost space station, yet preserve its essential research capabilities, and match those options to specific cost limits. The administration appointed an advisory committee to serve as enforcers to ensure that the redesign effort met these guidelines; the committee would forward its recommended option to the president. Additionally, the president himself warned that it would be necessary to redesign the space agency to accomplish this effort effectively.[54]

NASA found itself in the uncomfortable but necessary posture of having to both adapt its concept of the space station to imposed guidance and pass its work through the filter of an external jury. Having been the initiator and definer, guiding others to understand what a space station should be, it now faced a dramatic role reversal. Furthermore, it had to "sell" the new concept and maintain a coalition whose support was precarious, all the while streamlining and improving its own management. Besides reducing the size and cost, NASA pared back the space station occupancy period from thirty to ten years and set the crew size at four. Nine years after President Reagan's approval to proceed, it seemed that this might well be NASA's last chance to achieve its space station goal. Even James Beggs, the champion of the space station who as NASA administrator had won over President Reagan, sensed it slipping away. As habitability is always the most costly element in space systems, it was clear that a permanent human presence was in jeopardy.[55]

Over the course of three months in the spring of 1993, NASA completed the required space station redesign and presented three different concepts to the advisory committee, which then passed its recommendation to the president. Two options retained elements of *Freedom* to derive value from the $9 billion already spent on its development. The third bore no resemblance to *Freedom.* All versions were smaller, less capable, more austere concepts for a human presence in space than *Freedom* had been. Each option represented different tradeoffs between cost, size, complexity, schedule, number of assembly missions, ease of assembly and maintenance, and research capacity—and none was a perfect solution. The science task group on the redesign team concluded that each of the permanent presence options had adequate research capabilities, though degraded compared with *Freedom.*[56]

President Clinton then directed NASA to proceed with one of the simplified, scaled-down versions of *Freedom* as an investment in the future and America's technical leadership. By fall 1993 the administration and Congress agreed to a new international space station concept—now informally called *Alpha,* although several other names were under consideration. In their announcements, both the White House and NASA framed the new foreign policy role of the space station as a symbol of peace and cooperation for the post–Cold War era. Administrator Goldin varnished the redesign effort with familiar rhetoric: NASA was given a seemingly impossible task but achieved it, and the president's decision to continue the space station was "another important step forward on the space frontier." He coined a new description, as well: "The space station will be a knowledge engine on the high frontier."[57]

As the redesign effort had escalated into a political battle royale, a move in the House of Representatives to terminate the space station program almost resulted in a "none-of-the-above" choice; the measure failed by a single vote, grim evidence of the erosion of political support. Despite further calls to cancel the space station, the *Alpha* concept survived for the next several years, but with significant losses. Parts of *Freedom* survived, but there would be no companion observatories, no satellite servicing capability, and no depot or transportation hub. Research would focus on microgravity studies in just two fields—materials science and life science. *Freedom* was transformed into International Space Station *Alpha,* and eventually *Alpha* underwent another restructuring with further losses in 2001. Its final name became simply International Space Station, or ISS.

During the 1990s, both NASA and the media had to address the paradox of proceeding with a compromised station that could offer only a diminished human presence in space compared with earlier visions. Even so, the station continued to be depicted as extravagant or worthless; one such unfavorable view was an editorial cartoon of it as a black hole sucking money away from earth. What NASA called a "more efficient" design, the media called "scaled back" and "cut-rate." To counter the view that a reduced space station was not worth building, NASA expanded its efforts to reach the public with new messages about the potential of the (downsized) space station for research that would lead to practical benefits relevant, for example, to aging and women's health. The agency also armed its advocates with lists of reasons for use in defending the space station's value and correcting misperceptions.[58]

The fate of the habitation (hab) module represented the contradictions, functional and rhetorical, in the myriad space station design changes from 1986 through 2001. The architecture and meaning of a permanent human presence in space converged in the residence. NASA invited journalists to tour the full-scale engineering mockup of the space station so they could understand how crews would live and work in orbit, and the hab module crew quarters always drew interest as the heart or soul of the facility. More than the laboratories, this module would be the arena for post–Cold War and multicultural cooperation in spaceflight. Ironically, the element that literally housed the human presence in space was expendable; it gradually shrank from two modules for a crew of twelve to one module for a crew of eight to no habitat module at all. In the final downsizing, accommodations for six people were shoehorned into the work-spaces. A most fundamental, even iconic, element for a human presence—a "home sweet home"—disappeared in sacrifice to cost pressures.[59]

For two decades, both the architecture and the rationale for the space station suffered from NASA's self-inflicted management problems and externally imposed funding limits. The agency tried to adapt its messages to changing circumstances and to be resilient in response to crisis. None of this escaped the attention of news reporters who tracked the station's woes, editorialists who parried every change in message with a rhetorical challenge, and cartoonists who unflinchingly skewered the space station. One of the most cynical cartoons appeared in 2006. It showed the last piece of the International Space Station being installed and inscribed "This useless facility is now complete and can therefore be abandoned to build an equally useless facility on the Moon." This was not the only indication that NASA's perpetual campaign failed to win over those who saw no purpose for a space station. That perception defied resolution.[60]

Even John Noble Wilford, who had written eloquently about the space shuttle and the romance of spaceflight, became disillusioned with the space station. He expressed skepticism about the value of a large manned orbiting station and uncertainty that its promised benefits would materialize. As early as 1988, he sensed unease about the ability of the United States to maintain its presence in space, given the political and technical difficulties already evident in the space station effort. During the tumultuous period of redesigns, he wrote of a perception that the space program was like "a giant spaceship in a decaying orbit: Still capable of some brilliant flourishes, but directionless and drifting lower with diminishing ground support." Wilford, too, like the anonymous editorialists, began to long for a human presence with a noble purpose in space.[61]

As the first crew launched to occupy the International Space Station in late 2000, its most strident foe grudgingly acknowledged that there might prove to be some value in it. The international collaboration alone was a worthy diplomatic achievement, but many of the claims for the station had been "hyped beyond reason." The still skeptical *New York Times* would be watching for management and costs to be brought under control so the station might indeed become a world-class research facility. Those prospects seemed to dim in the aftermath of the 2003 *Columbia* tragedy, two different NASA administrators, and President Bush's new vision for space exploration, and by 2005, the *Times'* modicum of hope had evaporated. Its editorials again questioned the white elephant, now viewed as an albatross, that precluded more worthy human activity in space: "If this nation is to continue a human space flight program it makes sense to pick a more exciting destination than a space station circling endlessly in low Earth orbit."[62]

The perpetual, dogged campaign for a space station damaged NASA's credibility. The agency that could put men on the moon within a decade spent almost two decades muddling through design after design with ever-increasing cost estimates to finish a job that kept getting smaller. Critics questioned the very existence of NASA and speculated that it had so lost its way that it should be disbanded. Even Administrator Goldin, who presided over the shrinking space station, wondered whether the agency could regain its reputation and relevance. To many observers, the space station became a symbol of confusion if not failure, of bureaucracy grown stale and ineffectual. The case for a permanent human presence in earth orbit barely registered with the public; after ten and fifteen years of occupancy, many Americans were not aware of the International Space Station or the fact that American astronauts continued to live and work in space after the shuttles retired. An organization effectively framing its message would never slide that low in the regard of opinion-makers or leave the public that unaware of its flagship program. Nor would it lose faith with its own members, many of whom tired of the shuttle and station and were eager to move on to "real" exploration somewhere else in space. The *New York Times,* among others, consistently beat the drum for a more exciting, more adventurous space program, not necessarily predicated on human spaceflight.[63]

With the widespread internet access and social media use of the early 2000s, the space agency tried new ways to communicate with the public about space station operations. NASA's website became the portal for anyone interested to "meet" the crews, follow onboard activities, and learn about in-orbit research. Related websites and cross-links increasingly made it possible for people to explore the scientific facilities and experiments in depth. Crewmembers posted blogs and tweeted about living in space, reaching millions of followers via Facebook, Twitter, and other social media platforms. As smartphones and tablets proliferated, NASA developed mobile applications (apps) that enabled people to tune in to the International Space Station and to know when it would pass overhead so they could watch for it. NASA partnered again with IMAX to create *Space Station 3D* (released in 2002), using the new IMAX3D camera on the shuttle and space station to give audiences on earth an even more vivid experience of spaceflight. Televised "downlinks" connected the International Space Station crews with classrooms, and public appearances by astronauts back from duty on the station raised awareness. By the time the space station was completed, NASA was finally able to step out of campaign mode and into new forms of public outreach and engagement. Through these channels, the International Space Station could almost speak for itself directly to the public, and some of the negative attitude of the campaign years faded.

There was one more scare, however. As the space shuttle program wound down into retirement, its mission to be fulfilled upon completion of the space station, attention turned as always to cutting the costs of human spaceflight and redirecting NASA to new priorities. Just as the International Space Station neared a targeted completion date of 2010, a new administration under President Barack Obama began to consider closing it down in 2015. A five-year return on a twenty-five-year investment made little sense to the international partners, the space agency, and the scientific users. Instead, the space station got a reprieve to operate until at least 2020 (and possibly until 2024 or 2028) under the banner, again, of a "new era."[64]

Despite all the trials, the construction and assembly of the International Space Station, completed in 2010–11, was an impressive international technological achievement. Its final configuration spanned the size of a U.S. football field and had eight habitable laboratory and storage modules with accommodations for six people (plate 11). Through 2016 (and sixteen years of continuous occupation), fifty expedition crews had lived and worked there, and more than two hundred people total had been on board. At that time, the space station partners anticipated about ten more years in operation.

Space historians and policy analysts have expressed wonder that this space station survived its long turmoil, crediting "the adeptness of NASA at marshaling support . . . and maintaining a coalition of intensely interested parties that ensured their positions were heard" in the political arena. NASA had to promise to do more with less in a risky "spiral of overselling." While international commitments and the economic pressure of thousands of space station jobs helped keep the program from being canceled, NASA still did a good enough job mustering arguments, creating meaning, and holding on to key support to save the station more than once.[65]

As the space shuttle program ended and the human presence in space continued year after year on the space station, people inside NASA and beyond turned their attention to what would come next. Would there be another "next logical step"? A new cycle of visioning began to frame the future of human spaceflight.

Chapter 6 Plans: Envisioning the Future in Space

Episodically throughout the thirty-year space shuttle era, NASA or an external advisory group released a report on the future of human spaceflight. These slim paperbacks are artifacts of the way influential people in the shuttle era thought about the future of human spaceflight. They also contain a panoply of verbal and visual imagery that give form and appeal to the idea. Having won approval to take "the next logical step" to a space station in earth orbit, NASA and other spaceflight advocates turned their attention to constructing yet another imaginary. What might "a permanent human presence in space" become and mean? As the space station no longer was the goal to be achieved, it could now serve as the means to a broader end. Those entrusted with envisioning the future set to work to define what the next goal could be and describe ways to reach it. In the process, they massaged words and images to create visions meant to motivate and inspire NASA and the public. These often stirred controversy.

There was scant precedent for the blue-ribbon panels convened for the purpose of future-planning during the 1980s and beyond. In the 1960s, when NASA's mission was well-defined and the challenges were daunting—to win the space race to the moon—there was little pressure to start planning the future.

Many within the space agency assumed that the future was assured; they would return to the elements of the 1959 *Long Range Plan of the National Aeronautics and Space Administration* and build upon their capabilities to enable missions to the moon. That first plan had called for a human spaceflight program that included a permanent near-earth space station. Well before the lunar landings were accomplished, NASA began in-house studies toward that probable next goal, which could be viewed as a precursor for an eventual return to the moon to set up a base or as a springboard to Mars. It was not necessary to convene advisory panels to plot the future.[1]

President Eisenhower did request that an external body assess NASA's man-in-space plans in 1960, when lunar missions already loomed beyond the in-progress Mercury flight series. Although this activity was about evaluating, not planning, it set a pattern followed in later external reviews. The president, vice president, or president's science adviser would appoint a select group of distinguished members of the academic, corporate, and military communities to serve on special assignment for a few months to review NASA's overall mission or the direction of a particular program and provide some recommendations or conclusions. These reviews served as reality checks to give the White House confidence that NASA was not running too far afield or to reveal issues that the agency might not have brought forward. Such reviews could also give the White House cover for either endorsing or reining in the space agency's plans. Human spaceflight, then called "manned spaceflight" or "man in space," usually received careful attention. It was, in fact, the focus of the 1960 review, the first of this type.[2]

A similar ad hoc committee reviewed military and civil space programs for President-Elect Kennedy to brief him on their status. Chaired by Massachusetts Institute of Technology professor Jerome Wiesner, who was already advising both President Eisenhower and the new president on science and technology, this review panel accepted the inevitability of manned spaceflight but doubted the scientific or technical need for it and thought its importance was exaggerated in the public mind. They argued that space exploration could be carried out well by unmanned means, and the scientific accomplishments of unmanned missions far outpaced any that might accrue by men in space. Their report cautioned about the risk to national prestige of failure to launch or return a man safely in spaceflight. Advising the president who would within months make the boldest decision ever to advance human spaceflight, the group recommended that he consider canceling the man-in-space program, or at least minimizing its significance compared with other space activities.[3]

Newly inaugurated President Nixon convened the first VIP body explicitly tasked to address America's *future* in space. Within the scope of a total program of space science, exploration, applications, and technology, the group laid out ambitious options for human spaceflight, all of them seemingly extravagant in a constrained economic environment with flagging public interest in spending for space. Their unchallenged assumption was that continued human spaceflight after the end of the Apollo era was imperative. The report to the president drew from NASA's own report to the task group, and the result basically reordered the goals laid out in NASA's 1959 long-range plan, itself largely a reflection of what has been called the von Braun paradigm of the early 1950s.[4]

The essential elements were a space station served by a space shuttle, an earth-orbiting space station, a lunar base, a manned mission to Mars, and beyond that, exploration in the solar system. The rationale for these projects was incremental progress toward a growing human presence in space. Past achievements were but "a beginning to the long-term exploration and use of space by man . . . from the initial opening of this frontier to its exploitation for the benefit of mankind, and ultimately to the opening of new regions of space to access by man."[5]

These three reports document the rare occasions during the 1960s and 1970s that external reviewers formally engaged in deliberations on the vision, value, and future of human spaceflight. There were, of course, many technical studies of particular projects—Apollo applications and Skylab, the space shuttle, even the space station—but no comprehensive outside examinations of the purpose and direction of human spaceflight.

Internally, at mid-decade, a NASA task group carried out a review in response to President Johnson's request for space priorities beyond the 1960s. This effort produced a seventy-page bound and illustrated volume that showed how present capabilities could evolve into future programs. Manned operations was one of six categories covered with a view toward short- and long-term goals. The near-term focus for human spaceflight was extended-duration earth-orbital missions in modified Apollo command and lunar modules, essentially treating the spacecraft as mini–space stations for missions lasting up to ninety days. These would lead in the longer term to manned orbiting research labs. The primary rationales for extended missions and space laboratories were to study the effects of the space environment on the human body and refine space operations techniques, but the report included a long list of other scientific and technology research projects. Lunar polar orbit and fourteen-day lunar surface stays also were contemplated, leading eventually to a lunar base and exploration of the planets, starting with Mars. Although the motive for human spaceflight through the

Apollo missions was primarily political, NASA was developing a rationale for a future of manned scientific research and extended space exploration.[6]

An internal future-planning report appeared in 1976 as the Apollo and Skylab programs ended and activity escalated toward the shuttle. The Outlook for Space study group, made up of twenty participants from around the agency, saw it as their mission to look toward the twenty-five-year horizon and suggest how the space program could be of public service—that is, to identify projects that had the greatest potential to benefit people and society. They took a pragmatic approach, constrained by what was technically feasible and what responded to a definable human need. With little rhetorical flourish or call for greatness, they addressed only the basic question "why should this be done?" Of the twelve broad themes and 125 specific projects outlined in the report, not one was human spaceflight itself, although one of the chapters focused on exploring and developing the space frontier. If humans were needed to meet a particular practical goal in space, they would be included, but human spaceflight per se was not a goal.[7]

The *Outlook for Space* report grouped needs into physical (improved health, food, medical care, energy, environment) and intellectual (exploration, discovery, knowledge, education). Earth-oriented projects would address physical needs, and extraterrestrial-oriented projects would address intellectual needs. "Living and Working in Space" fell into the category "use of the space environment for scientific and commercial purposes" in response to human desires for exploration and knowledge. Some of these needs might be met by exploiting a human presence in a space station in low earth orbit or eventually exploring beyond, but these were only broad brush statements, not blueprints for a human presence in space.

In general, the *Outlook* report was rather dry and staid, in keeping with the practical emphasis on the space shuttle in the political and economic environment of the mid-1970s. It presented not so much a vision of the future as a menu of non-controversial options that would connect the space program to basic human needs—the kind of down-to-earth approach that a pragmatic president like Gerald Ford or Jimmy Carter might have warmed to. It did not make the pulse race with excitement.

In contrast to this earlier period, external and internal reviews occurred frequently during the shuttle years and played a role in NASA's continuing mission to establish a permanent human presence in space. Beginning in the early 1980s, when the shuttle seemed to be achieving routine spaceflight with increasingly frequent missions, both NASA and panels of distinguished citizens strove to articulate increasingly compelling versions of the imperative for human spaceflight. At least twenty-five significant reports appeared in as many years, some more memorable than others, but each an effort to frame a vision for a permanent human presence in space. The new era in space also became a new era in planning.

What was different in the shuttle era, and why was so much effort expended to scope out the future of human spaceflight? Did the vision substantially change, or was the same rationale simply reframed in different guises? More important, did these studies and reports have any genuine influence on decision-making and the course of events? Both the context and content of envisioning the human future in space from about 1985 onward bear analysis.

At least two situational conditions distinguished the thirty-year shuttle era from the Apollo era. First, the space program and NASA were no longer deemed as critical to the national welfare as during the Cold War competition of the 1960s. After the Apollo successes, NASA had to clamor for its place at the table with other agencies and priorities when the White House and Congress negotiated budgets, and it found itself needing to stoke an ongoing public relations campaign to keep its mission before the public and decision-makers. Without strong enough political patronage for its choice programs, NASA had to make its case continuously, year by year, budget cycle by budget cycle. Studies and reports that addressed a future filled with promise were an important part of its survival as a significant presence on the national agenda.

The second distinctive difference was the advent of strategic planning in the public sector and its spread in the 1980s and 1990s. Strategic planning had an ancient history in the military arena, but it did not become a trend elsewhere until the twentieth century, first in city planning, then in business and industry, next in government, and then everywhere—in educational and religious institutions, nonprofit organizations, and even in personal self-help and career coaching. Strategic planning exploded into a popular growth industry, as evidenced by the thousands of books and consultants dedicated to its practice. Its advent in government agencies occurred just about the time in the mid- to late 1980s that NASA and others began to look seriously at the human future in space beyond the shuttle and space station. By the early 1990s, strategic planning was becoming so ingrained in organizational management that Congress passed a law requiring all federal agencies to develop and operate by a strategic plan. Soon, all agencies would implement performance measurement based on metrics linked to their mission statement and progress in meeting their strategic plan's goals and objectives.[8]

Although strategic planning became a popular trend, there is nothing magic about it. It is an orderly process of deciding what or where an organization wants to be in the future and charting the path(s) it will take to get there. It is a success-oriented discipline that helps an organization turn its goals into realities. Integral to strategic planning are creating a vision, aligning the vision with the organization's beliefs and values (its culture), and convincingly communicating the vision to all stakeholders. The shared vision then drives all decision-making and action plans, and it is the basis for commitment and unity of purpose.[9]

Strategic planning has much in common with the communications technique of framing ideas to influence opinion and win support, as earlier discussed. It, too, is a deliberate exercise in the construction of meaning and definition of success. It, too, must be rooted in shared beliefs and values to be credible and appealing. At its most mundane, strategic planning produces a roadmap to a goal; at its most sublime, it inspires and motivates.

The crucial foundation of strategic planning is creating a vision, or visioning, or visualizing a goal. Each form of the term means seeing. The strategic vision is not simply an idea or an intellectual concept. To become compelling, it must be something that its supporters can see as real. For that reason, the vision is often shaped into a narrative or an image that makes it coherent and believable. It may also be incarnated in a logo or brand that serves as its symbol.

Visioning and leadership are often linked, because it typically is an organization's top managers who guide strategic planning efforts; in the case of blue-ribbon strategy panels, all the members are leaders in their professions. According to leadership expert Warren G. Bennis, "Leaders articulate and define . . . then they invent images, metaphors, and models that provide a focus for new attention. . . . An essential factor in leadership is the capacity to influence and organize meaning for the members of the organization."[10] Others concur that "the vision begins with a guiding idea, but the strategist must understand and master the art of the metaphor.[11] Strategic planning may well yield a new ideology, rhetoric, and iconography for the organization's mission.

From the 1980s onward, NASA itself, as well as appointed panels outside the agency, engaged in envisioning the future. They published strategic plans loaded with images and metaphors devised to present a tantalizing yet credible (*achievable* was the term of choice) view of human spaceflight in the future. Of the many future plans and reviews from this period, it is possible to derive the major themes and directions from selected documents, especially pairs of internal and external reports published about the same time. The rhetorical frameworks of these reports reveal much about the mindsets and rationales of the planners who attempted to craft the meaning for a permanent human presence in space.

Several times during the shuttle era, NASA found itself at a pause point, forced by events to consider altering its course to the future of human spaceflight. The 1986 *Challenger* tragedy prompted the first such pause as pundits, politicians, and some within the agency pointedly questioned the wisdom of the spaceflight enterprise and an investigative commission probed the space agency's culture and practices. Another occurred at the end of the 1980s, with the space station in the throes of redesign, when the first President Bush announced a surprising charge to go to the moon and Mars. The space station vision dominated the 1990s, but the 2003 *Columbia* tragedy triggered another pause to reconsider the purpose of human spaceflight and the shuttle. By 2010, with the end of the shuttle program in sight, yet another study ensued.

Each of these events could have been a turning point or crossroads, an opportunity to make a dramatic change in direction by re-visioning the goal of a human presence in space. Each time, leaders proclaimed it was time for a new sense of purpose or new long-range goals. Instead, each new study or plan served as a threshold from which to step out toward a reframed past goal. Once building a space station was approved, visionaries could finally be explicit about the long-understated goal of establishing a human presence in space beyond earth orbit. Human settlement of the solar system—at least the moon and Mars—moved to the top of the agenda, becoming the vision that drove strategic planning for the nation's space program.

At each of these moments of pause, two planning documents emerged—one written by an appointed commission and another by a NASA study group. The future they described and recommended was remarkably consistent over twenty-five years: expanding the human presence in space and maintaining U.S. leadership there. The studies presented variations on the same themes; they repeated key ideas unchallenged, mixed in familiar metaphors, and used visual images that played to long-held expectations. Despite significant energy poured into visualizing the human settlement of space on a timeline starting in the mid-1990s to 2000, each successive study pushed that horizon back to the early and then the mid-twenty-first century. These strategic planning efforts were eloquent but ineffectual in launching progress toward the envisioned goal, because they inevitably met political obstacles and assault by critics of human spaceflight. Reports meant to be inspirational too often appeared with a flourish but ended up shelved in the archives. Ultimately, the perpetual exercise in creating a vision for the human future in space yielded little that was new and failed to turn the old ideas quickly into reality. That future remains uncertain to this day.

The foundation report for those that followed appeared in 1986, by coincidence just four months after the *Challenger* tragedy. It was the product of the National Commission on Space, chartered by Congress and appointed by President Reagan. The impetus for this commission arose in Congress in 1983, shortly after the space shuttle was declared operational and well before the loss of *Challenger*. As NASA launched its headlong campaign for a space station, enough members of Congress were concerned about the direction of the space program to adopt legislation requiring a citizens commission empowered to "study existing and proposed United States space activities; formulate an agenda for the civilian space program; and identify long range goals, opportunities, and policy options for civilian space activity." Said an influential member of the House of Representatives, "America had a space program headed in all directions and no direction. There was a failure of vision in regard to our future in space."[12]

NASA protested that such a commission was unnecessary and redundant to its own planning authority as well as the prerogatives of the administration and Congress. Nevertheless, Congress levied the requirement and set aside $1 million in the NASA budget to fund a one-year study. President Reagan issued the executive order to establish the National Commission on Space, named its fifteen members, and appointed former NASA administrator and aerospace executive Thomas O. Paine to serve as its chairman. The members were renowned leaders in government, academia, business, and the military; they included two astronauts and two women; and they ranged in age from seventy-five to thirty-four, with an average age of fifty-six years, meaning that most were mature during the 1960s "golden age" in space. The group, which came to be known as the Paine Commission, had twelve months, later extended to eighteen months, to submit a report that fulfilled its charter.[13]

Pioneering the Space Frontier: An Exciting Vision of Our Next Fifty Years in Space created a stir upon its release in May 1986 (plate 12). With colorful, futuristic cover art and illustrations by noted space artists, the report unabashedly endorsed "exploring, prospecting, and settling the solar system" as a major goal. Adopting the frontier narrative that had appealed to President Kennedy at the dawn of the space age and inspired their present patron, President Reagan, the panel liberally framed the future in terms of that idealized past. They also returned to the past vision of Wernher von Braun, acknowledging that current human activity in space realized much of what he had foreseen in the 1950s. The report, with its vision of humans on Mars, was both an electrifying antidote to the post-*Challenger* gloom and conspicuously naïve in its spirited optimism about complex space endeavors.[14]

Within the pioneering framework, the National Commission on Space crafted goals for the future in metaphors of construction—building a technology base, a highway to space, and a bridge between worlds. It included spaceports, fleets of transport vehicles, outposts, resource mines, and processing plants—a civil engineering utopia that had a contemporary ring and also a suggestion of the work of opening the frontier West. The vision was so expansive that it embraced the spread of civilization "from the highlands of the Moon to the plains of Mars" to make the solar system the home of humanity. Evidently the commission was comfortable enough with the pioneering-frontier conceit that the members did not consider another approach to framing the future.[15]

Although their charter directed them to focus on the next twenty years in space, the commission chairman decided to look to the fifty-year horizon, knowing how long it takes for ideas to come to fruition. The cover of the report, which was released by a New York publishing house and sold in bookstores, featured a dramatic scene of a thriving settlement on Mars. Drawing attention at the center of the scene, perched high above the human activity, stood a space-suited figure in the iconic stance of an explorer—staff in hand, arm raised in a salute, the (U.S.) flag in view—looking like Columbus in the New World, now the master of Mars. This image alone telegraphed that the report would present "an exciting vision of our next fifty years in space." That vision was based on analysis, not just dreaming; the commission laid out many technical and economic prerequisites of a phased approach for settling on the moon and Mars. Their report argued that the bold vision of the future was achievable if the nation willed it so. It did not pass unnoticed that it was an expensive, even extravagant, vision of the future.

Toward that end, the commission recommended that NASA strengthen its planning activity to ensure that the future receive sustained attention. NASA welcomed the report and then initiated a future-planning activity of its own, under the leadership of astronaut Sally Ride in the newly created Office of Exploration. She had come to Washington after serving on two shuttle missions and on the Presidential Commission on the Space Shuttle Challenger Accident. Interested in helping NASA get back on track after its trauma, she put together a task group of representatives from the NASA centers for a year-long strategic planning study. In August 1987, her group published *Leadership and America's Future in Space*.[16]

The "Ride report" was a more focused blueprint than the Paine Commission report, as if the study members had rolled up their sleeves to make a credible start toward a grand vision. To their credit, the group wrestled with the term "leadership" instead of just plugging it into their title. "Leadership cannot simply be proclaimed," they wrote; "it must be earned." Their metric for leadership at a meta-level was presence in space—either the presence of automated explorers to other worlds or the presence of human beings. Leadership meant having capabilities, actively using them for sound purposes, and achieving visible, significant results. From such leadership would flow inspiration, national pride, and international prestige. These latter were the product, not the aim, of leadership.

They also described leadership as a process of strategic planning and execution, an integral part of which was communication—articulating, promoting, and defending goals. The report modeled the first step in planning by identifying and evaluating four distinct initiatives to establish leadership in space, two of them an expanded human presence in exploration. It modeled clear-eyed thinking about goals and the means to achieve them. It offered fresh but not grandiose concepts, and it was restrained and reasonable, without hyperbole. An example of such clarity was a simple statement about a hot-button topic: "Settling Mars should be our eventual goal, but it should not be our next goal." That goal should not be hurried. A time-critical goal received this forthright explanation: a Mission to Planet Earth initiative "is a great one, not because it offers tremendous excitement and adventure, but because of its fundamental importance to humanity's future on the planet." Ride's group strove to distinguish between hype and authentic merit in goal setting.[17]

These two reports, appearing within a year of each other, were in some ways polar opposites in their framing of the future. The *Pioneering* report was exciting in its bold vision of making the solar system humanity's home, its stylish publication, and its optimism that great ambitions were possible. The *Leadership* report was calm and temperate, a modestly designed document with a conversational tone, prepared by a group likely affected by the *Challenger* tragedy and Sally Ride's service on the investigative committee. Its message was reassuring: America's leadership in space could be regained by diligent effort. In contrast, the *Pioneering* report was so ambitious as to seem a fantasy, despite its thoughtfulness. Yet both were earnest efforts to envision an achievable future in space.

One is tempted to wonder how the leadership styles of Thomas Paine and Sally Ride may have influenced the approaches of their teams and reports. Did it matter that he was a senior executive and she was an astronaut and national icon, that he was an older male and she a younger woman, he a businessman and she a scientist, he appointed by the president and she a NASA employee, that the commission was composed of celebrities supported by a staff and a million-dollar budget while the Ride group comprised NASA's in-house talent? The reports alone do not suggest answers, but a study of records of both teams' work would no doubt shed some light on the influence of the leaders' personalities.

The main difference in the two reports was the way they framed the future. The National Commission on Space report's title gives it all away: *Pioneering the Space Frontier*. The future would fit into the mold of America's heritage. The artwork was new but the idea was old; it had been President Kennedy's theme twenty-five years earlier. Thomas Paine ardently promoted vigorous space exploration as an expression of America's pioneering spirit, and upon leaving NASA, he tirelessly promoted an expansive view of space exploration. The National Commission on Space membership did not include a historian. Its members were scientists or leaders in government and industry. Would pioneering on the frontier have been adopted as the conceptual framework had a Western historian or author—perhaps Patricia Nelson Limerick, Wallace Stegner, or another scholar with deep knowledge of the real frontier experience—served on the commission and opened a much different angle on American history? We cannot know, but the power of this metaphor in the twentieth century was such that it resonated from generation to generation and was easily adapted for many purposes. The frontier had particular resonance in the first generation of space program participants and observers who had come of age in the 1950s when Westerns were popular on television and at the movie theater, and when American history textbooks glorified the opening of the West. Time and again, planners brought the frontier into service as the framework for thinking about space exploration and human spaceflight.[18]

The *Leadership* report, in contrast, did not wear this familiar paradigm. It framed the future as a product of responsible human behavior based not on "a vision" but on "clear vision." In fact, this little gem was almost buried in the preface to the report: "A clear vision provides a framework for current and future programs." The argument was that leadership in space depends on clarity in goals, strategies, and execution. The words "pioneering" and "frontier" rarely appeared in the leadership report; the theme of clarity provided its structure.[19]

What the two reports had in common was the effort to address a constant concern in the political and news media arenas that the United States space program was adrift, without a clear purpose or direction. Even though NASA had two major human spaceflight programs (shuttle and space station) in progress in the mid-1980s and had a family of great observatories and planetary explorers soon to be launched, there was a widespread sense outside the agency that none of this was coordinated as parts of a master strategy. It seemed like a piecemeal approach to space exploration, with all the programs in competition, some suffering in deference to others, and all of them subject to continual rethinking and reconfiguring. Both the Paine Commission and the Ride reports attempted to bring order and sequence to space exploration, sensing rightly that a haphazard approach could not sustain public and political support.

NASA's usual nemesis, the editorial board of the *New York Times,* had for several years battered the space agency for lack of vision, and its responses to these planning activities was not encouraging. Its only comment on the Paine Commission's report was that a mission to Mars, if carried out jointly with the Soviet Union, might be worthwhile, but it faulted the government for not giving NASA clear goals and priorities, and it faulted NASA for being too enamored of human spaceflight. A scathing editorial near the first anniversary of the *Challenger* tragedy denigrated "the poverty of NASA's dreams." It claimed that "NASA's musty plans and cramped vision lack both utility and imagination . . . the space program has become a yawn." In its view, NASA's goal should be to restore adventure and discovery through vigorous exploration of the universe by automated and robotic means.[20]

In early 1987, the *Times* published a feature article by John Noble Wilford on the allure of Mars as a national goal in space. He sensed the momentum coming from the National Commission and Ride studies but noted that it would probably be very difficult to launch another multibillion-dollar project while the space station was mired in controversy. However, such a goal might give focus to the agency and also be the exciting challenge that the present space program seemed to lack. The *Times* editorial board was keen on a joint American-Soviet program of Mars exploration, and the fact that media attention to the two reports always highlighted Mars exploration suggested its greater appeal.[21]

During the months after the *Challenger* tragedy, the space policy community sensed that NASA and the United States were at a juncture for rethinking the direction of the space program and gaining some clarity about its fundamental purpose and goals. In this climate, the *Times* ran a guest opinion piece by an academic who dared to challenge the space frontier rhetoric. "NASA's selling of space as a frontier is a con job," he stated, after explaining how space is nothing like a frontier. He noted that the rhetoric sounds good because the frontier is important in the public imagination; it resonates with adventure and greatness, but the reality of space is much different. This writer viewed the Ride report as the latest evidence of NASA's refusal to grow up and be responsible in pursuing a realistic, affordable space program. He thought such a program was possible only by giving up the frontier metaphor, which "intoxicates politicians and bamboozles the public . . . excuses failures and exaggerates promises."[22]

Whether or not it was framed as a pioneering venture on a new frontier, the potential goal of Mars exploration stimulated discussion, or at least invited distraction from current problems. Space policy pundits were pleased to comment on it as a plausible remedy for the space program's malaise or as a misguided plunge into another morass. NASA's standing independent advisory council endorsed the exploration of Mars as a primary goal that would capture broad public support and enhance U.S. technical and scientific capabilities. Of course, Mars had been in the long-range plan since 1959 (and imagined as a destination before formal planning), so it was not a new goal. But the timing of its reemergence in 1986–87 was significant. The space shuttle was temporarily grounded, and the space station was still an unsettled concept. It may have been the best of times to present an exciting vision of the future, but it was the worst of times for it to gain any traction as a real program. Nevertheless, expanded exploration in the solar system, especially Mars, by both machines and humans, was now explicitly on the agenda and would remain so for at least the next thirty years. All strategic planning efforts that followed the National Commission on Space and Sally Ride's study group would look toward Mars.[23]

The two reports had another virtue in common: both stressed the importance of public engagement. The National Commission on Space put that into practice by inviting public input through open forums around the country, surveys, and letters from citizens. A chapter of *Pioneering the Space Frontier* dwelled on the dual flow of communication—education of the public and listening to the public—necessary to gain support for an ambitious program to explore and settle the solar system. This report recommended many ways to make space exploration more visible, and for NASA to explain to the public better what it does and why. The *Leadership* report briefly addressed the need for an informed public prepared for decision-making about national goals and a well-educated population to do the difficult technical work of space exploration. The essential purpose of framing the meaning of spaceflight, or any other cause, is to engage the public by making it intelligible, appealing, and worthy of commitment. Both study teams understood this.

As to the fate of their reports, after the initial splash, they were evidently shelved—not immediately, but eventually. Neither received a prompt formal response from NASA or the White House beyond gracious acknowledgments of receipt and compliments on the effort, and concerns arose that the reports would be ignored.[24] The law required NASA to review the National Commission's report and respond with an implementation plan. The agency demurred that it would be "premature and presumptuous" to do so beyond reaffirming its commitment to the goal of expanding the human presence in space.[25] When Sally Ride announced her departure from NASA months before her report was due, rumors circulated that a possible reason was frustration over lack of high-level interest in the planning effort her group had carried out so diligently. She stayed at the agency until the report was released and then moved on to make a difference elsewhere. The new NASA Office of Exploration she headed was later reorganized out of existence.[26] Neither report became a blueprint that was consulted to guide the next steps of planning. Each new leader at the top of NASA and in the White House wanted to put his own stamp on the future and set a new agenda.

On July 20, 1989, a new president stood on stage at the Smithsonian National Air and Space Museum with the Apollo 11 crew and other astronauts and dignitaries to mark the twentieth anniversary of the first landing on the moon. President George H. W. Bush seized the occasion to make a major space policy announcement—his vision for human spaceflight. The United States would build space station *Freedom*, then return to the moon, and then go on to Mars to extend a permanent human presence well beyond earth. He delivered the speech with touches of humor, nostalgia, and pride in past achievements. He made passing mention of explorers and the frontier, but his rhetorical framework built on two different foundations. One was concrete; he spoke of the pathway, stepping stones, and milestones on this course into the future. The other was intangible; he spoke of the dream, opportunity, challenge, and commitment of becoming the world's preeminent spacefaring nation. His message echoed that of the National Commission on Space, whose report had been prepared while Bush was Reagan's vice president: "The time has come to look beyond brief encounters. We must commit ourselves anew to a sustained program of manned exploration of the solar system and, yes, the permanent settlement of space." And finally, he answered the question "Why?"—because it was human destiny to explore and America's destiny to lead.[27]

The Bush proposal became known first as the "Human Exploration Initiative" and then the "Space Exploration Initiative." It was a clear but skeletal conceptual framework for charting the future in space. After years of complaints among political and media figures, some of them NASA's friends, that the space program had no clear focus, goal, or direction, President Bush gave it not one but three. He defined where to head and tasked the National Space Council, chaired by Vice President Dan Quayle, to define when and how. Members of Congress caught by surprise criticized the lack of specifics and funding to bolster the president's vision. But for a couple of years this initiative put human spaceflight goals on the horizon and gave parts of the space community a purpose stated at the highest level of government. President Bush's speech gave hope to those who believed that another Kennedy-style presidential mandate was the ticket to the future. Before that hope faded, good work happened.[28]

NASA made the first response to the president's speech. The new administrator, Admiral Richard Truly, formed a task group led by Johnson Space Center director Aaron Cohen to prepare a report for the National Space Council's use. Intended to be a reference, not a set of recommendations, the report assessed current capabilities and the need for advances in order to meet the president's three goals over a thirty-year period. It also gave several different approaches to meeting those goals as examples for the council's use in strategic planning. To make the point that the report was not a sales pitch for any particular program, Cohen called it a database to aid policy makers; it was a starting point for discussion, not the final word. The modest-looking report was prepared in-house in just ninety days. Although not a fancy publication, it did something noticeably different. It did not use the standard concept of the frontier as its rationale. This report eloquently introduced a variant for framing human spaceflight.[29]

Seemingly out of nowhere, and with no reference cited, NASA's ninety-day study report advanced the idea of "the exploration imperative." It is worth extracting a few sentences that link this vaguely biological concept with American values and the human future in space.

> The rationale for exploring and settling space mirrors the spirit that has compelled explorers through the ages: the human urge to expand the frontiers of knowledge and understanding and the frontiers where humans live and work. . . . The imperative to explore is embedded in our history, our traditions, and our national character. . . . The exploration imperative propels us toward new discoveries.
>
> Exploration is a human imperative, one deeply rooted in American history and its destiny. . . . Exploration is an endeavor in which our Nation excels. Returning to the Moon and journeying to Mars are goals worthy of our heritage, signaling an America with the vision, courage, and skills essential for leadership among spacefaring nations in the 21st Century.[30]

Where did this idea originate that was stated with such certainty? Western thought embraces a few imperatives—biological, territorial, moral, categorical—but is there a validated concept of exploration as an imperative, or is this only a rhetorical device? The prose style, although not the meaning, was reminiscent of Carl Sagan's. His wildly popular *Cosmos* television series and book in the 1980s had graced science and space topics with poetic language. The idea of an exploration imperative was reminiscent of Wernher von Braun's belief that movement into space was inevitable, and of President Kennedy's insistence that the nation not be left behind as others set sail on the ocean of space. It also sounded somewhat like Neil Armstrong's explanation of the reason for going to the moon: "because it's in the nature of the human being . . . it's by the nature of his deep inner soul. We're required to do these things just as salmon swim upstream." It is an evocative assertion, but whether it has a scientific basis is debatable.[31]

The president's Space Exploration Initiative and NASA's response triggered related studies in 1990–91 by the National Research Council, an advisory committee appointed by the White House, a think tank, another appointed study group, the Office of Technology Assessment, and others for consideration by the National Space Council. Each review or planning group was made up of experts with different perspectives who met occasionally, led portions of the study, reviewed drafts, and debated their findings until reaching a publishable consensus. Some explicitly called their report a framework for analysis, but in a sense all were shuffling a deck of possible options and dealing them according to their own rules. As President Bush had given open-ended guidance without schedule or cost targets, the variety of scenarios for returning to the moon and advancing to Mars proliferated. (Bush later suggested a thirty-year timeframe aimed at a U.S. human landing on Mars by 2019, to mark the fiftieth anniversary of the Apollo 11 landing.)[32]

Far more text was written about "architectures," "infrastructure," and "configurations" than a rationale for the human exploration initiative. Attention to the reasons for establishing a human presence on the moon and Mars varied from study to study, but writers generally subordinated "why" to options for "how" and "when." The National Research Council report, written by a committee of fourteen senior experts (none women) in science, engineering, and management, confined its analysis to technical and scientific capabilities. This group did not address the "exploration imperative" or any other rationale except to state that scientific research should not be considered the primary justification for human exploration of space.[33]

That *Human Exploration of Space* report did not circulate widely in public; its primary audience was NASA and the National Space Council. A much higher profile report came from the Advisory Committee on the Future of the U.S. Space Program appointed by the White House to conduct a thorough review of the civil space program, including management issues. The director of the National Space Council at the time made it clear that the committee should do a top-down review of NASA itself, then plagued with shuttle, space station, and

Hubble Space Telescope problems. The NASA administrator was not happy with this charter, and word leaked to the media that the survival of NASA was at stake.[34]

Martin Marietta chairman and chief executive officer Norman Augustine led this twelve-member committee that included two members of the 1986 National Commission on Space. Co-chair Laurel Wilkening—scientist, academic, and the lone woman—also had co-chaired the earlier commission, and member Thomas Paine had chaired it. The circle of those whose service was desirable for "blue-ribbon" space panels evidently was small enough that some were tapped again. Although committee members came from different professional zones—industry, government, military, academia—the diversity within these bodies was quite limited, if their purpose truly was to bring new ideas and fresh perspectives. Most brought managerial competence to the discussion, but most were also insiders already steeped in the status quo and perhaps not primed for taking a fresh look at space exploration.

What came to be called the Augustine Commission or Augustine Committee report published at the end of 1990 addressed concerns about the prospects for planning and executing an ambitious human exploration initiative. Among them were lack of a developing consensus on how to proceed, NASA's overcommitment and management turbulence, institutional aging and technology stagnation—all of which contributed to a resistance to new ideas and change. Although the report alluded to the president's goals for space exploration, it sounded lukewarm about a rationale based on such intangibles as "the desire to explore." Its framework for the future was to balance a mission *to* planet earth and a mission *from* planet earth on the fulcrum of science, depicted exactly so in a simple drawing. In their view, "science gives vision, imagination, and direction to the space program." For once, the *New York Times* responded favorably, agreeing with the panel that science was the proper goal for reorienting the space program and urging Congress and the administration to support that priority.[35]

This advisory committee's report recommended that because of myriad uncertainties in such a long-range goal, further human exploration should proceed in step with available funding—that is, on a pay-as-you-go basis. Although it endorsed the goal of expanding human presence to the moon and Mars, absent the pressure of a competitor in the post–Cold War world, this report essentially put brakes on the human exploration initiative. It punted the initiative downfield in favor of solving current problems and revitalizing a culture of achievement within the troubled space agency. This made it easy for NASA to ignore the report and did not give the National Space Council a strong basis for a policy decision during the one-term Bush administration. Wrote the council director years later, "Plans for an American renaissance in space exploration to pivot, not pause, at the end of the Cold War, were bogging down."[36]

One other study group from that period deserves mention for its novel approach to assessing the Space Exploration Initiative for the National Space Council. NASA, the RAND Corporation, and a number of government agencies and professional organizations conducted a widespread appeal for ideas. Former astronaut and executive Thomas Stafford led a group that reviewed this input and "synthesized" it into a document titled *America at the Threshold: America's Space Exploration Initiative*, often called the Stafford report or Synthesis Group report (plate 13). Like *Pioneering the Space Frontier*, it was a colorful, glossy well-illustrated book designed to capture attention. Of the several efforts mentioned here, this was the only one that explicitly addressed the question "why?" Citing examples from the fifteenth century onward, the report opened with the caution that "nations lose their leadership position when they give up the role of exploration." The goal of expanding human presence in space posed the challenge and held the promise of world leadership in science and technology. This report had a more optimistic outlook than the others discussed here. It recommended that the Space Exploration Initiative become the centerpiece for future planning and that it be buttressed with strong education and outreach programs that would both prepare the public to participate and engage the public to support the long commitment. However, the report presented a framework of multiple visions and architectures that might shape the future in space, thereby complicating the picture.[37]

As more and more studies were called for and undertaken, so many options and architectures appeared that the Space Exploration Initiative receded from possibility. In all versions, the cost estimates annually and spread over time looked prohibitive. Opposition coalesced in Congress to deny startup funding, and there was no advocacy coalition strong enough to advance the cause. The only emerging consensus was that it would be premature to embark on a detailed plan to reach the moon and Mars in a foreseeable timeframe. President Bush's broad vision simply entailed too many technical and cost unknowns to be seriously pursued in a time of an enormous national deficit and budgetary strictures. A human presence in space beyond a space station that was still years from being started could simply be dismissed as unrealistic and unaffordable.

It was not clear that any substantive progress occurred before the Space Exploration Initiative collapsed in late 1992. By then, President Bill Clinton and a new NASA administrator were on the way in. Historians have judged the Space Exploration Initiative a failure as a proposed program and as a policy process but a success as an "idea generator," for it stimulated widespread thinking about planning approaches, technologies, management systems, and other topics. However, the flurry of intense study activity did not produce agreement on a compelling motive for human exploration.[38]

Each study group presented its own version, typically a variation on those presented five years earlier by the National Commission on Space and Sally Ride's task group: the legacy of pioneering or the call to leadership. Space exploration meant "destiny"—humanity's and America's. An idea emergent from these studies held some promise as a unifying rationale: that science, not politics or prestige or other peripheral goals, should drive exploration. This enabled a renewed focus on robotic exploration and fed dialogues about human-robotic partnership in space. Thinking about exploration in the framework of science permitted a metric for decisions about human presence, whether it was the only way or best way to gain desired knowledge. The more original idea of a semi-biological/semi-cultural "exploration imperative" did not catch on as the framework for journeys and settlements beyond earth, unless perhaps "imperative" and "destiny" shared the same meaning.

Instead, the focus in the 1990s became the International Space Station and integration of Russia into the program after the collapse of the Soviet Union. With pressure mounting to build the station, NASA and the White House dropped planning for future human spaceflight beyond earth orbit, although robotic exploration activity began to increase. From 1997 onward, robotic landers and orbital observatories made their way to Mars as science missions. Whatever they learned would be valuable for planning any future human missions, but human journeys to the moon and Mars were off the agenda until the International Space Station was assembled, occupied, and fully operational. That would not happen until the first decade of the new century.

In parallel with these special reports in response to President Bush's initiative, NASA had an internal strategic planning process that aimed to clarify the mission of the agency at its broadest and align its many activities with that overarching mission. Administrator Richard Truly issued one of the early editions of the annual plan in January 1992. Called *Vision 21* (for the twenty-first century), it fit the pattern of stating the agency's values and principles, vision and goals, and strategies for achievement. As a roadmap to the future and leadership plan, it included a mission from planet earth to expand human activity into the solar system in accord with the president's challenge and recommendations from the various studies discussed above. In flat prose, it seemed not a particularly inspired prospect. The plan's morale-boosting language about maintaining leadership and achieving excellence were out of phase with the troubles NASA was experiencing in its marquee projects. The shuttle still had hiccups, the Hubble Space Telescope needed repair, the space station was mired in controversy, and the Space Exploration Initiative was fading. Neither the present nor the future looked as promising as the strategic plan implied. Such disparity between vision and reality did not correlate with efforts to frame the meaning of human spaceflight for the next century.[39]

With a full schedule of human spaceflight in earth orbit geared toward the purpose of establishing a permanent human presence on the space station, NASA's longer-range planning activities received little attention for several years. NASA's 1995 reorganization into "enterprises" patterned on a business model resulted in the creation of the Human Exploration and Development of Space (HEDS) enterprise entity after the Office of Exploration was disbanded. In the 1990s HEDS became the main enclave for thinking about the future of human spaceflight in the new century. According to NASA's strategic plan for 1995, the HEDS mission was to "open the space frontier by exploring, using, and enabling the development of space . . . to bring the frontier of space fully within the sphere of human activity." The mission sounded grander than the reality, because functionally its programs were limited to earth orbit. However, it also engaged in studies for the evolution of human spaceflight beyond the shuttle and space station.[40]

The pace changed when NASA found itself in another forced pause, this time after the tragic loss of *Columbia* and crew in 2003. With the shuttle fleet grounded, space station assembly halted, and an accident investigation in progress, the future of NASA and human spaceflight was again in dispute. Would both be able to survive the second deadly blow, despite the past fifteen years of relatively routine successful missions? How might human spaceflight be framed against the sudden reminder of its inherent risk? Would discovery that the *Columbia* loss mirrored some of the problems that had doomed *Challenger* kill the shuttle program?

The younger President Bush stepped forward in early 2004 to offer a new vision predicated on the resumption of spaceflight, completion of the International Space Station, and retirement of the shuttles by 2010. He proposed as the next steps the development of the first new crew spacecraft in a quarter century and its use in earth orbit by 2014 and for a return to the moon in 2015 to 2020. Without committing to Mars, he acknowledged that the moon was a step along the way to farther journeys. (The exploring rover *Spirit* had just landed on Mars the previous week.) While giving due credit to robotic explorers, he noted that people are not satisfied until they can see and touch for themselves. He then introduced the chair of an advisory committee to help NASA determine how to carry out this plan and encouraged him to get to work right away.[41]

The president quickly framed the new vision for exploration as a renewal of the spirit of discovery. Like Lewis and Clark, he said, we venture into space because we have the desire to discover and understand, because it is in our character, and it brings us benefits that improve our lives. Echoing a phrase from the space policies of the Reagan and previous Bush eras, he said it was time to "extend a human presence across our solar system." The new era of discovery would be a long-term commitment, a journey, not a race, a choice that "improves our lives, and lifts our national spirit."

As a framework for human spaceflight, this announcement of a vision for exploration had more structure than that of the first President Bush. This one included specifics—a timeline, a technology project (the Crew Exploration Vehicle), a directive to NASA to align its activities accordingly, and an initial funding plan. It had elements to disarm potential critics—a bigger role for robotics, a link between future funding and achievement, a marathon pace instead of a sprint, a sustainable commitment. It also used exploration rhetoric with nuance, touching on several familiar themes without forcing the vision into any one of them. Knowledge, risk, advanced technology, innovation, research, benefits, discovery, courage, inspiration, progress, challenges, opportunities, journey, national spirit—all these keywords were brought into service to form what this President Bush expected to be a "great and unifying mission." This vision had everything the former Space Exploration Initiative announcement lacked.

The timing for this renewed vision might have been just about right, to judge by some influential media commentary. The outlook had begun to turn after the *Columbia* accident. Almost immediately a *New York Times* editorial on space exploration was atypically positive about human spaceflight; despite the cost and risk, it was important to maintain and refocus spaceflight on the "ultimate if distant goal: human exploration of far-off worlds." Why? Because humans are compelled to explore, wired for curiosity and knowledge, making it inevitable that humans will eventually set foot on another world. The piece ended with this encouragement: "The task ahead is to find a way to keep the flame of exploration alive at a time when the space program has been rocked by tragedy." That perspective appeared again a few months later in an editorial urging a national debate on the issue of continuing human spaceflight in earth orbit or reaching for "bolder, more adventurous missions deeper into our solar system. . . . Sooner or later, humans ought to venture once again to distant worlds."[42]

Closer to the White House, the *Washington Post* expressed skepticism that such an "audacious goal" was appropriate in a deficit environment and that

NASA could break its cost overrun habits to succeed with a fiscally sustainable program. A satirical editorial cartoon showed a wheelchair-bound girl reading the news and asking, "They are prepared to spend how much so a man can walk on Mars?" However, the hometown newspaper for NASA and the president did not attack the new vision for exploration, so the climate seemed to be softening somewhat, if only because the post-*Columbia* hiatus in spaceflight demanded some kind of plan.[43]

NASA was ready. In February after President Bush's January speech, NASA published a well-illustrated twenty-eight-page booklet called *The Vision for Space Exploration,* "a new, bolder framework for exploring our solar system." Its rationale was primarily science and discovery, and its method would be "sustainable, affordable, and flexible." The vision integrated robotic and human exploration into a combined strategy, and it committed to a return to the moon in preparation for possible future human exploration of Mars and other destinations. It would focus research on the International Space Station to support exploration goals, primarily understanding the effects of long-duration spaceflight on human health. The initial technology development would be a new spacecraft for crew transportation to the space station or the moon. And it would massage NASA's organization for better coordination in fulfilling the vision. It was an invigorating vision.[44]

At last, the agenda for planetary science, human spaceflight, technology development, and the space station were focused on common goals in a synchronized strategic plan. Finally, the rationale was solid, perhaps less assailable, with its sensible emphasis on science and discovery. Rhetorical uses of the frontier or leadership were subordinated to new primary themes—curiosity about the unknown, and the appetite for knowledge, growth, and change. This vision fit a science and engineering endeavor without pretenses of grandeur. Accomplishing the vision would effectively bring leadership, prestige, inspiration, and the other lures. NASA distributed the booklet to congressional offices, the media, and the space exploration community and set about the effort of persuasion.

At the same time, the advisory commission led by former secretary of the air force Edward "Pete" Aldridge settled into its 120-day study to recommend how to implement the vision. The charter for the nine-member commission (which included three women and a prominent African American man) was to consider the science, technology, organization, and management needed to accomplish the goals, as well as ways to engage the public, private sector, and international community. In due course, they issued a colorful sixty-page report endorsing the vision. A stylized logo graced each page—three circles along a curvy line, representing the path from earth orbit "to the Moon, Mars, and Beyond"—a graphic icon for the associated slogan.[45]

The report held few surprises, but it emphasized the core problem of sustaining a political commitment to a necessarily multi-decade program of exploration. Success would depend on a series of presidential administrations, congressional sessions, and NASA leaders working in concert despite changes in personnel and the economy. It would depend on the public remaining committed as well. Success would depend on better discipline to operate in a "go as you can afford to pay" mode. It would also be necessary to transform certain cultural obstacles, such as aversion to risk and intolerance of failure, if technology breakthroughs were to occur. The recommendations of the report addressed those and other requirements under the "imperatives" of sustainability, affordability, and credibility.

The Aldridge Commission made no attempt to reframe the vision rhetorically; President Bush had done that clearly enough as a renewal of the spirit of discovery arising from the pursuit of knowledge. In their view, the open questions were how, not why. They accepted the premise that "the impulse to explore the unknown is a human imperative" and elaborated on many benefits that could be anticipated from the program of journeys. Their contribution was a set of findings and recommendations commensurate with the framework that the president had established.

NASA organized a new Exploration Systems Mission Directorate to carry out the vision and gave the new flagship program a name: Constellation. A new logo appeared, featuring three small orbs to represent the earth, moon, and Mars. Planning focused on two new launchers, named Ares I and Ares V in allusion to Mars. The crew exploration vehicle, an Apollo-type capsule sized to hold six people, became Orion. Constellation's slogan became "Moon, Mars, and Beyond," both a literal goal and a sly echo of the newly iconic popular film character Buzz Lightyear's signature statement, "To infinity . . . and beyond!" Work progressed from 2005 on but struggled with "inconsistent and unreliable" funding that stretched out the schedule and increased costs.[46]

Soon after taking office in 2009, President Barack Obama decided to appoint an independent review panel to evaluate the current state of human spaceflight plans and make recommendations to ensure the nation was on "a vigorous and sustainable path to achieve its boldest aspirations in space." He appointed a committee led by businessman Norman Augustine, who had also chaired the Committee on the Future of the U.S. Space Program in 1990. Upon doing their due diligence, the group determined that the Constellation program could not be achieved at its authorized funding level; a much greater infusion of funds would be required for a meaningful human spaceflight program. This conclusion proved to be the kiss of death despite five years of spending already committed to developing the program. Intentionally or not, the title of the report—*Seeking a Human Spaceflight Program Worthy of a Great Nation*—insinuated that past and present human activity in space was unworthy. NASA morale plummeted, and editorial cartoonists caught the irony. The unpalatable report did introduce some new ideas into the conversation: that of options and "a flexible path" to destinations as funds and technology permitted, and a mission to an asteroid as part of the "Moon, Mars, and Beyond" thrust (plate 14).[47]

President Obama canceled Constellation in early 2010 and directed NASA to concentrate on spurring new technologies for spaceflight. Some of the technical legacy from Constellation might carry over into the development of a new crew vehicle and space launch system, but a return to the moon was off the agenda. The president acknowledged the importance of human spaceflight for advances in knowledge and technology, but he wanted NASA to find a smarter, more affordable, and sustainable way to do it. He also wanted to shift responsibility for providing transportation in low earth orbit to the commercial sector. In canceling Constellation, Obama did not establish a vision or framework for human spaceflight, effectively leaving it in limbo without a plan and shifting emphasis from "where" and "when" to "how." The administration and the NASA administrator spun this abrupt change of plans as a "bold initiative" to invent new approaches to human spaceflight, but the media recognized it as a huge setback, possibly portending the end of human exploration in space. *Florida Today*'s editorial cartoonist captured the frustration NASA surely felt by depicting Obama as Buzz Lightyear pointing to space and proclaiming not "To infinity . . . and beyond!" but "To rather vague yet somewhat achievable possibilities and beyond!" Typically sympathetic to NASA, the same artist also acknowledged the agency's need for a forward direction (fig. 6.1).[48]

The demise of both the Space Exploration Initiative in 1992 and the Vision for Space Exploration's Constellation program in 2010 stalled the momentum of planning for human spaceflight. Expansion of human presence in the solar system would not begin anytime soon after the space shuttles ceased flying. In both instances, it was not so much the rationale that failed, but the funding and politics. Even as Constellation died, many stakeholders could agree on the role of human spaceflight for purposes of exploration and discovery. After some twenty-five years of strategic plans and reviews, a viable consensus on "why" seemed to be settled, but the "how" and" when" were in disarray. It was time to return to the drawing board.

Fig. 6.1. Despite a plethora of future plans during the shuttle era, NASA faced continued criticism for lacking a vision. Jeff Parker for *Florida Today*, Reprinted by permission of Cagle Cartoons, Inc.

Strategic planning for the future of human spaceflight punctuated the shuttle era. While thinking beyond the present endeavor was essential for institutional survival, it was also a healthy exercise in creativity. As future planners worked to organize their ideas, they also tried different ways of framing them for persuasive appeal. They mined the frontier metaphor, or adopted images from architecture and civil engineering, or promoted leadership to convey what human spaceflight could mean. Presidential initiatives stimulated intense rounds of creative work to generate scenarios and "architectures" for a future human presence in space. Going to Mars and settling the solar system—ideas that appeared in the earliest NASA planning documents—officially gained acceptance, albeit temporary, keeping the von Braun paradigm of methodically expanding human activity in space alive into the new century.[49]

A few trends emerge from these vision-creating and strategic planning efforts. One is the increasing sophistication of communications evolving from simple typescripts with diagrams to glossy color publications, from roundtable discussions among experts to open public hearings with citizens' input, and from static written materials to interactive websites. Visual rhetoric gained importance in reports that looked like magazines illustrated with carefully selected photographs and images, sometimes with original art commissioned for that purpose. Marketing came into play in the packaging of plans and reports for visual appeal to gain attention and be memorable. Inevitably these plans would stir up controversy, so NASA distributed campaign-style lapel pins and other mementos with related logos to spread the message and show support. Sometimes these trinkets became the only physical manifestation of programs that died without being realized.

The process of planning for human spaceflight became more open with the awareness that the public is the primary stakeholder. That is not to say that it became more inclusive. Most of the designated commissions were fairly homogeneous in membership, composed primarily of male senior executives from technical organizations. The blue-ribbon panels were insular to the extent that the members always came from other bureaucracies with ties to NASA; there were no true outsiders who might bring genuinely fresh perspectives and challenges to the deliberations. Members were eminently qualified to review technical and managerial issues, but few if any had primary expertise in the rhetoric of public communication. Occasionally, a study group brought in a historian as a consultant to offer background information more than fresh ideas for the future. Editorial support staffs for the commissions usually handled the actual composition of commission reports. The enlistment of primarily engineering and scientific minds may have been a limiting factor in producing a persuasively original framework for the future of human spaceflight. Public hearings around the country and broad solicitation of ideas via letters or the internet mitigated this homogeneity to a degree by opening channels for the commissions to receive other input and take the pulse of public attitudes toward spaceflight.

Ultimately, envisioning the future in space starts with three simple questions: where do we want to be, why, and how do we get there? Answering those questions calls for imagination tempered by realism, or realism stretched by imagination, and the answers must be credible. They must resonate with beliefs and values outside the plan itself. The technical credibility of NASA's visions for the future has not been the primary issue; cost and schedule have been the main points of contention, and to a lesser extent the motivating rationale. The problem has been that the rationale has never been widely enough embraced to work through the cost and schedule issues; without a commitment to the rationale, it has been too easy to simply ignore planning reports or cancel initiatives. This suggests a basic weakness, if not failure, in framing the meaning of human spaceflight.

NASA's plans for human exploration have always issued from the engineering parts of the organization, where the natural strength is dealing with technologies. Engineers excel in understanding the realities of human spaceflight and creating hardware and software to operate in that "real world." But the "why" question is not one to be answered in engineering units alone. For the goals of exploration and discovery, the science parts of the organization have obvious expertise to identify research priorities and determine which require human presence. Yet scientific objectives are typically very specific and hard to relate to an exploration rationale that is intelligible to nonscientists. Answers to the "why question" from scientific units alone also may not satisfy. Planning groups tasked with envisioning the future in space could benefit from having some members or advisers who are historians, communications professionals, policy analysts, creative young entrepreneurs, and professional futurists such as creators of space science fiction. Such members would bring fresh perspectives and make valuable contributions to framing plans more resonant with the public whose sustained support is required. The high-profile failure of plans for future spaceflight initiatives strongly suggests that new approaches to rationale development are needed.

In January 2011, NASA released an in-house study, the Human Space Exploration Framework Summary, that matched President Obama's expectations after the 2009 review. Unlike previous studies, this one did not target a destination as its focus, although its cover art still depicted the moon, Mars, and an asteroid. This was a framework only for developing capabilities (technologies) that would enable any of several future missions, with destinations to be determined later when new launch and crew systems became available. It offered no rhetorical rationale, just four criteria for all elements: safe, affordable, sustainable, and credible. Written in dense management jargon that could be almost unintelligible to the uninitiated, the report nevertheless had little nuggets of freshness, such as the question "Can multiple paths get us where we want to go?" illustrated by a multicolor subway system map, and a "capabilities catalog" from which missions could be assembled. The engineering and management analysis may have been solid, but the "packaging" lacked those inspirational flourishes and allusions that graced other such reports. This one was businesslike and bureaucratic, unintentionally making the effort to prepare for the future sound like drudgery.[50]

At the time of this book's publication, NASA was marching along a "flexible path" outlined in the 2009 critique and 2011 framework. But to invigorate that workaday approach, the agency had unveiled an inspirational phrase for its future: "America's Next Giant Leap." This new slogan appeared in July 2014 in time for the forty-fifth anniversary of Apollo 11 as a title over the iconic image of a boot print on the moon. However, in this image the gray lunar soil transitions to orange Martian soil (plate 15). This phrase and others—"the human path to Mars" and "Journey to Mars"—became hallmarks of speeches, publicity materials, banners on NASA's website, and its social media outreach, all seeking to generate the excitement of the time when human spaceflight was a bold adventure.[51]

Meanwhile, as shuttle and space station era astronauts retired, NASA recruited a new class (the twentieth, in 2009) to prepare for future exploration. They may be among the first to fly in the new Orion spacecraft and possibly venture beyond earth orbit some fifty years after humans last visited the moon. NASA has also invited the public to join in studying Mars and thinking about best landing sites and research projects there. Active public engagement, especially with the millennial and younger generations, became an energetic strategy condensed into one oft-repeated statement: "The first humans who will step foot on Mars are walking the Earth today."

The appropriation of imagery and words from a past golden age is either an ingenious or desperate rhetorical ploy for envisioning the future. It remains to be seen whether it lights a fire of public and political support that will enable a human future in space. Meanwhile, the vision of a grand future in space exists largely in the artifacts and memories held by museums.

Chapter 7 Memory: Preserving Meaning

NASA chose April 12, 2011, as an auspicious date to make the announcement. It was the thirtieth anniversary of the first space shuttle mission and the fiftieth anniversary of the first human spaceflight by cosmonaut Yuri Gagarin. NASA administrator Charles Bolden, a four-time shuttle veteran as pilot or mission commander, addressed a crowd in the bright Florida morning outside the orbiter processing facility at Kennedy Space Center. The aft end of *Atlantis* towered behind the stage, flanked by the flags of the United States and each of the shuttle orbiters and five large "We Made History" banners, each bearing the name of one orbiter. This backdrop, accented by the NASA, United Space Alliance, and shuttle thirty-year logos, was a picture-perfect distillation of Bolden's message about "this tremendous American accomplishment" that was nearing its end.[1]

As he paid tribute to the workforce for their long commitment to the shuttle program, viewers watching the NASA television broadcast from space museums around the United States waited anxiously. When he reached the main point of the occasion, Bolden briefly teared up and paused, as he often did in moments of emotion. "The shuttle's retirement is bittersweet for us, but . . . our commitment to human spaceflight is steadfast. We will continue to lead the world in human

Fig. 7.1. For many in the shuttle workforce and the public, seeing these "birds" retire to museums was a bittersweet end to decades of spaceflight, yet many appreciated that the orbiters and their stories would be preserved. Courtesy of Ed Stein. Reprinted with permission.

space exploration and discovery—and don't let anyone tell you otherwise." He then proceeded to announce where each of the orbiters would be placed on permanent display in museums. With the end of the space shuttle program in sight, it was time to preserve its legacy for present and future generations (fig. 7.1).

Displaying the orbiters and telling their stories offer new opportunities in the half-century effort to articulate the meaning of human spaceflight. For the shuttle era, it is no longer presidents, NASA, advisory committees, and the media that have the lead in shaping public understanding of its rationale; the framers of meaning now, as for the space-race era, are historians, museum curators, and educators. As stewards of the iconic spacecraft and their narratives, it is their responsibility to interpret the shuttle era for the public. Forthcoming scholarly and popular accounts may well introduce fresh metaphors and new interpretations of familiar symbols, leavened with critical judgments clarified by hindsight.

The end of the space shuttle program had a long prelude. Within a year of the 2003 loss of *Columbia* and crew while returning from a successful laboratory science mission, President George W. Bush announced the inevitable. Speaking at NASA Headquarters in Washington, D.C., he stated that the shuttle would be retired upon completed assembly of the International Space Station, expected in 2010. The shuttle had done its job, and there were no future missions to justify the expense of maintaining the fleet. NASA would be refocused to develop a new space transportation system for exploration beyond earth orbit. Ahead lay a called-for return to the moon by 2020 and preparations for an eventual human journey to Mars. In setting this new course, the president affirmed familiar values from the history of spaceflight—"daring, discipline, ingenuity"—and entrusted the future to the "risk takers and visionaries of this agency." He even resorted to the "logical step" rhetoric from the 1980s space station campaign as he hailed a "new era of discovery."[2]

For NASA, the president's vision for space exploration was a double-edged mandate. It gave new goals, direction, and reassurance that human spaceflight would continue despite the latest tragedy. But the plan also signaled a coming hiatus in U.S. space transportation capability and a diminished capacity in the future. Despite its vulnerabilities, the space shuttle was a huge cargo vehicle that transported more people, equipment, and supplies than any other craft leaving the earth. The primary values that had propelled the shuttle into space—routine, practical spaceflight in a reusable multipurpose vehicle—would be abandoned. The proposed new mode of space travel had a déjà-vu aura—an expendable crew capsule launched on top of a rocket for "back to the future" forays to the moon. Deemed safer and more economical than the shuttle, it seemed more like a step backward than forward.

News analysts, editorialists, and editorial cartoons greeted this new vision with skepticism. It was but heady rhetoric without an infusion of funds, a point made more than once by *Florida Today* cartoonist Jeff Parker, whose newspaper served communities affected by the vagaries of NASA funding. The *Washington Post* carried Tom Toles's satirical view of the new space policy in a caricature of Bush at a podium with a poster depicting the moon, Mars, and piles of money titled "Nothing for Something." The *New York Times* reported concerns over the billions in likely costs and trying to do such an ambitious program "on the cheap." As usual, the debatable question about the so-called moon-Mars initiative was "why?" The president offered that humans are never satisfied in the quest for knowledge unless they see and touch for themselves, but critics charged that the program would perpetuate human spaceflight without a clear advantage of people over advanced robotic explorers. This was a timely complaint as the intrepid *Spirit* and *Opportunity* rovers began their explorations of Mars to wide acclaim.[3]

Media attention to the Bush vision for space exploration dimmed as the shuttle's return to flight approached in 2005. Editorial cartoons served as either harbingers of the end of the space shuttle era or as slightly less confident affirmations of America's resilience. Uncle Sam watched the launch with his fingers crossed for good luck, and the supply checklist included duct tape and a rabbit's foot. One cartoonist depicted persistent delay-causing faults—loose tiles, faulty sensors, and other problems—in a tournament bracket. In this time of transition, old memes reappeared in new guises. The space truck was now an antique pickup or a vintage Cadillac, obsolete and needing repair.[4]

By 2006, a new meme appeared: the shuttle personified as a geriatric patient plagued with ailments. *Discovery,* the return-to-flight orbiter, was stooped and wheelchair-bound, connected to an intravenous line (fig. 7.2). Reuters ran a cartoon depicting a cane, walker, Geritol, and prune juice as "necessary safety

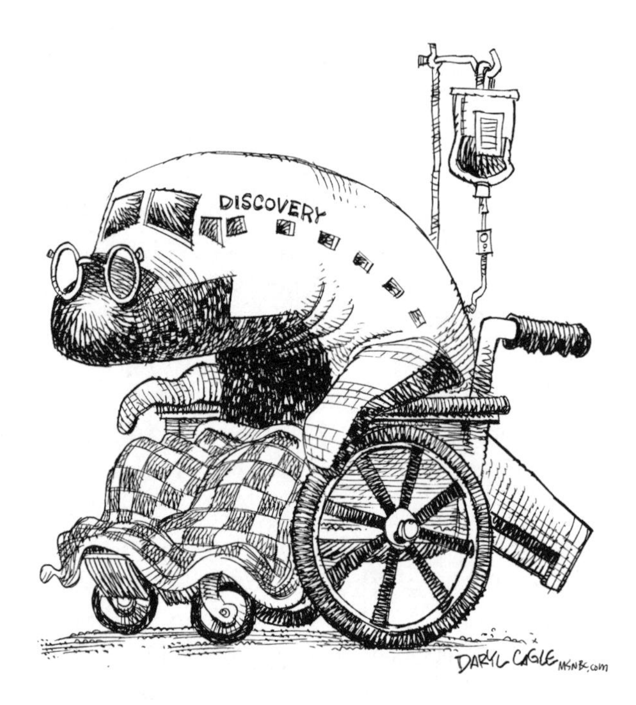

Fig. 7.2. After the 2003 loss of *Columbia* and crew, cartoonists depicted the space shuttle as old and decrepit, a perception that did not mesh with reality; the orbiters had been updated often enough that they were in many ways new vehicles. Daryl Cagle for MSNBC.com. Reprinted by permission of Cagle Cartoons, Inc.

measures." Even the *Orlando Sentinel* added to the indignities, depicting an incontinent shuttle shedding tiles as it toddled along behind a walker. Meanwhile, the actual shuttle soldiered on as the space workhorse it was meant to be, delivering massive piece after piece of the International Space Station plus tons of supplies, and ferrying crews up and down. The exaggerated gap between reality and such cartoonish perceptions fed the notion that the shuttle was an aged vehicle, more than ready for retirement, even though the end of the program had more to do with its cost than its performance.[5]

Meanwhile, NASA set about bringing the shuttle era to a close. In 2008 the agency issued its plan to transition assets out of the space shuttle program and retire the spacecraft, related property, and part of the workforce. It was a huge undertaking; beyond the orbiters, boosters, and tanks were scores of buildings and facilities, an estimated million or more pieces of equipment, and several thousand workers. NASA had to decide what real property to retain for further use in new programs, what to set aside as historic artifacts for preservation, and what to demolish and dispose of. Certainly the spacecraft elements and some body of to-be-specified spaceflight items would be designated as artifacts and made available to museums and educational institutions. At the retirement of the Apollo program, the Smithsonian National Air and Space Museum was the only well-established repository for such artifacts, and NASA sent truckloads of equipment there. By the time the shuttle program ended, a number of NASA visitor centers, community science centers, and aviation and space museums were thriving around the country; they were eager for an opportunity to collect artifacts from the space shuttle program. To prepare for the eventual release of property into this widespread community, NASA published a guide to the types of artifacts that might become available and an overview of the acquisition process.[6]

The most prized artifacts of the space shuttle era were the three remaining flown orbiters—*Discovery, Atlantis, Endeavour*—and the atmospheric test flight vehicle *Enterprise.* All previous U.S. spacecraft but two resided in the National Collection preserved by the Smithsonian, and most were displayed on loan in other museums to make them more widely accessible. It was generally assumed that a shuttle would land in the National Air and Space Museum, but it was impractical on several counts to consider assigning responsibility for all three flown shuttle orbiters to the Smithsonian, which already had custody of *Enterprise.* Because only NASA had the ability to transport orbiters, any placed on loan would in effect be left permanently in that location. It made better sense to transfer each orbiter directly into the ownership of its permanent home. To that end, in late 2008 and early 2010 NASA issued "requests for information," essentially soliciting applications from interested organizations that might qualify to receive an orbiter for display and preservation. NASA then managed an evaluation and selection process that became very competitive as entities seeking an orbiter mustered community and political support to bolster their case.[7]

After long months of anticipation and suspense, NASA Administrator Bolden announced the recipients on that April morning in Florida. *Atlantis* would stay in Florida, to be housed in the Kennedy Space Center Visitor Complex at the home base of shuttle operations. *Endeavour* would go to the California Science Center in Los Angeles, in the area where the shuttle was designed, developed, tested, upgraded, and often landed. The Smithsonian National Air and Space Museum would receive *Discovery* to be displayed in its space hangar in Chantilly, Virginia, near Dulles Airport and Washington, D.C. (fig. 7.3). The museum had requested to exchange *Enterprise* for *Discovery,* the longest-serving orbiter, as the one most suitable for the National Collection, but had no assurance from NASA until shortly before the announcement. And finally, *Enterprise* would move to the Intrepid Sea, Air & Space Museum in New York City. The

Fig. 7.3. Twenty years before the end of the space shuttle program, a northern Virginia artist foresaw that an orbiter would join the iconic air and space craft held by the National Air and Space Museum. This cartoon appeared upon the announcement that the museum would open a new center near Dulles International Airport. Used with permission. Mike Jenkins, © 2016.

decisions thrilled four museums and left a number of others unhappy, especially Space Center Houston, next door to the home of the shuttle program and the astronaut corps. Officials from some of the disappointed communities threatened to challenge NASA's choices, but the selection based on population and tourism in the chosen sites held up to legal scrutiny without contest.[8]

The orbiter assignments signaled the end of the program more tangibly than any other action in the multi-year phase-down. *Discovery* had just completed its final mission in March 2011; after the placement announcement, *Endeavour* and *Atlantis* flew their final missions to the International Space Station in May and July in a combination of work and victory laps. *Discovery* retired with thirty-nine missions to its credit; *Atlantis* had thirty-three, and *Endeavour* twenty-five. *Columbia*'s twenty-eight and *Challenger*'s ten brought the total number of space shuttle missions to 135 over thirty years. The process of releasing thousands of smaller artifacts from the shuttle program continued for several years until NASA's shuttle workshops and storage facilities were bare for use in other programs. Some workers transitioned to new programs, some retired, and several thousand lost their jobs as the shuttle era drew to a close. For them, the mood at the end was more somber than celebratory. As Bolden said, it was bittersweet to reach the end of a career spent in the spaceflight enterprise and face an uncertain future. The intense pace of spaceflight had been their way of life for so long that the coming hiatus was worrisomely reminiscent of the fleet groundings after the tragedies. Without a spacecraft in the hangar, it was not clear that there would be a return to flight.

NASA's delivery of the iconic orbiters to their new homes in 2012 signaled another transition: a passage into memory. A museum is a "memory institution" with a charter to preserve and interpret the artifacts and narratives of the past. It is a place where personal knowledge and memory often intersect with presentations of the past that scholars variously call public, social, cultural, collective, or communal memory. Such memory is not entirely factual; it is an interpretation of facts set in broader social and cultural contexts of meaning.[9]

All four vehicles went to institutions that celebrate technology, science, and their roles in history, but each has a distinctive mission and had to decide how to present its orbiter within an appropriate framework of meaning. What they did would shape public memory of the shuttle for years to come. The shuttle's history was recent enough that museum staff and their immediate audiences knew its story primarily through lived experience, yet those to come who did not remember the shuttle in action would need a meaningful narrative. How would these museums present the meaning of the space shuttle era? Would they devise a technological or social narrative, acknowledge changes in meaning over time, wrestle with contested meaning as expressed by the space shuttle's critics, or adopt some other approach?

These questions get to the heart of the history and museum professions and of recent scholarly literature that addresses who owns the past as presented in museums and monuments. Artifacts may be icons to those who know their provenance, but to others they are mute objects that demand explanation. Who is privileged to tell their story and shape public memory; who decides what will be remembered, included, and excluded? Interpretation of the past is one of the most contemplative responsibilities held by curators and historians, who are bound by standards of historical research and intellectual rigor. The staff responsible for curating the orbiters at each museum were passionate about spaceflight. Attaining some critical distance in telling a shuttle story would be a challenge as each institution chose how to present its orbiter as the icon for spaceflight in the shuttle era.[10]

Even before receiving the space shuttle orbiters, museums began to curate these artifacts; their applications to NASA presented preliminary ideas for display concepts and educational programs. The end of the shuttle program opened the possibility of expanded spaceflight exhibits to accommodate the anticipated release of thousands of related objects. Hopeful institutions made preparations during the two-year wait for NASA's orbiter placement decisions to be announced. They began to map out floor space and thematic treatments for a shuttle or other large items—training simulators or solid rocket booster segments, for example—and in several cases planned entire new buildings. After NASA announced the four orbiter locations in April 2011, the pace accelerated to become ready for 2012 deliveries.[11]

Most of these internal preparations were evident mainly to museum staff, with limited external awareness. Only when the orbiters arrived did the public begin to understand how the space shuttle era would be curated. Each recipient museum staged a grand public celebration to welcome the shuttle to its community. With differences in rituals at each location, the museums had a common motive: to greet their orbiter in a manner befitting a national icon.

Discovery arrived first, entering the National Collection of treasures maintained by the Smithsonian Institution. The National Air and Space Museum coordinated closely with NASA and other federal agencies to arrange for a public celebration worthy of the nation's capital. It began with a first sighting in airspace on the outskirts of Washington, then two low passes over the city. As the Boeing 747 shuttle carrier aircraft with *Discovery* mounted on top slowly flew a loop around the National Mall from the monuments to the Capitol and then followed the beltway to fly over the Maryland and Virginia suburbs, people poured out of museums, schools, shops, and office buildings to spot the shuttle. Excitement ran as high as if this were the debut, not the final flight, of the orbiter. The day's imagery captured *Discovery* with the Capitol, the Washington Monument, and other prominent national icons before its landing at Dulles International Airport in suburban Virginia. Images of this flyover appeared on all the major television networks that day, April 17, and more than one hundred newspaper front pages in the United States and abroad the next day.[12]

Two days later, after technicians de-mated the orbiter from the 747 aircraft and lowered it to the ground, the museum officially received *Discovery.* Given the federal status of the Smithsonian, the museum orchestrated a festive but dignified welcome ceremony around patriotic rituals and spaceflight symbolism. Recognizing the one hundredth anniversary of Marine Corps aviation (and the fact that both the director of the museum and the NASA administrator were retired USMC generals), the U.S. Marine Drum and Bugle Corps performed marches and military service hymns, and the U.S. Marine Color Guard presented the flags. Opera star Denyce Graves, a Washington native and resident, sang the National Anthem. Cheered on by a flag-waving crowd, some thirty astronauts in flight suits and a retinue of shuttle workers in their customary attire escorted *Discovery* to the area outside the museum's space hangar where *Enterprise* waited; the two orbiters—the oldest and the most flown—parked nose-to-nose as the backdrop for the ceremony. Waiting on stage for the astronauts were the NASA administrator, the chair of the Smithsonian Board of Regents, the secretary of the Smithsonian, the National Air and Space Museum director, and Senator John Glenn, who had returned to space on *Discovery.* Each spoke eloquently of the significance of the day and of the shuttle, and more broadly of American traditions of exploration and innovation. Upon their signing the official transfer document, *Discovery* became a national treasure to be held in trust in perpetuity.[13]

The National Air and Space Museum's charter is to collect, preserve, study, and exhibit artifacts of the nation's achievements in aviation and spaceflight. It presents the history, culture, and science of aviation, spaceflight, and space exploration in order to "Commemorate, Educate, Inspire." At a national museum chartered by Congress, and funded in part by taxpayers, the public expects a positive and uplifting presentation of the nation's accomplishments and heroes. That was the tenor of *Discovery*'s welcome.

Speaking at the arrival and welcome ceremonies, secretary of the Smithsonian Wayne Clough took the lead in expressing how the Smithsonian would frame the meaning of the shuttle era. He cited the institution's long history in aviation and space science research as a perfect match for the orbiter's history. Clough welcomed *Discovery* as "more than a marvel of engineering and technology." It also was "an engine of imagination, education, and inspiration" that communicates across cultures, a "reminder of the power of human ingenuity to solve great problems," "a challenge to heed the wisdom" that exploration is the essence of the human spirit, and a reminder that "what we did before, we can do again." Clough commented, "When I look at *Discovery,* I see the strength of our nation, the strength of its people." John Glenn also spoke of the significance of *Discovery,* starting with a brief litany of the nation's tradition of pioneering and innovation, and its successes in the effort to go "up there." Glenn thought of *Discovery* on display as a testament to the era, an inspiration for the future, and also "a symbol for our nation of space flight that represents optimism, hope, challenge, leadership and aspiration to explore and to excel."[14]

In style and substance, the ceremony occurred symbolically, if not physically, in the capital of the United States, and the space shuttle stood as a national icon of America's space-exploring spirit. At least 150 radio, television, and press organizations covered the events. An image or article about the transfer made front pages around the world, and prime-time radio and television news also gave global coverage. Web hits, Facebook, and Twitter posts skyrocketed. Nothing in recent shuttle history had captured so much favorable attention, prompting someone present to comment that if NASA had brought the shuttle out for public appreciation like this more often, political support for the program might have been stronger.[15]

After the ceremony, *Discovery* rolled into the center of the museum's space hangar and was on display to the public the next day as if it had just landed there. After some adjustments from its ferry-flight configuration for static display, the first phase of exhibitry appeared. *Discovery* occupied the central floor space where *Enterprise* formerly stood, in front of a large United States flag on the hangar wall. The orbiter's presentation there matched the display scheme for all the aircraft and spacecraft in the Udvar-Hazy Center—a forthright, unembellished statement of its significance in the history of flight. Instead of a single label panel allotted for each artifact, *Discovery* claimed four: one each to describe the orbiter's significance, its features, its propulsion system, and its record-setting history of thirty-nine missions in space. Its last launch played continuously on a video monitor to preserve the memory of *Discovery* in motion. The museum had asked NASA to leave *Discovery* as intact as possible so it could serve as the technical reference vehicle for future study. Displaying the orbiter on the floor in landing configuration did not require any alterations beyond those NASA made for safety reasons or to retain certain components, notably the three main engines, for other programs. In addition, the logistics of this display mode were simple and the cost negligible.

The museum placed *Discovery*'s Canadarm remote manipulator system beside it rather than leave it hidden from view inside the payload bay; such an instrumental feature of shuttle missions deserved to be seen. Three shuttle payloads were already suspended above *Discovery:* a manned maneuvering unit flown by astronauts away from the shuttle on three 1984 missions, a full-scale model of the tracking and data-relay communications satellites deployed from the shuttle, and a radar topography mapping instrument flown on a mission in 2000. One of the two flown Spacelab modules sat behind the orbiter (plate 16). This cluster of related artifacts highlighted some of the main achievements of the shuttle era. The museum would move more elements into place as they became available, including an interactive exhibit that enabled visitors to explore via touchscreen the interior flight deck, middeck, and payload bay as if they were inside the vehicle. The ability to put themselves virtually in the commander seat or float through the middeck into the payload bay created an almost palpable sense of being a shuttle astronaut.

As significant, across the aisle from *Discovery* the shuttle's precursors already stood on display: an unflown Mercury capsule that Alan Shepard would have flown in orbit had the mission not been canceled as unnecessary, the Gemini 7 capsule in which Frank Borman and James Lovell spent two weeks in earth orbit rehearsing for a lunar mission, and an Apollo training capsule decked out with the Apollo 11 flotation devices. The immediate comparison in size indicated what a quantum leap in spacecraft design the space shuttle was. Its presence on the floor, as if it had just landed, prompted many visitors to exclaim in surprise at how large the shuttle was and how impressive the spaceplane was compared with the capsules. Unlike the view on television and in photos, where the orbiter was diminished by distance and its towering tank and boosters, the face-to-face encounter and walk-around to see the full wingspan, tail height, and main engine nozzles made an indelible impression of magnitude. The proximity of these spacecraft also suggested a spaceflight lineage in which the shuttle was a noticeable departure and the most recent, but probably not the last, vehicle.

As the first orbiter to be retired, *Discovery* represented the end of the shuttle program, an occurrence that caught much of the public by surprise. Spaceflight had become so routine that it was in the background, and people who had not followed NASA news well were astonished that the shuttle era was coming to an end. Because the shuttle had become the icon for human spaceflight, many people (including some in the media) assumed incorrectly that the astronaut program also was ending. NASA had to keep explaining that the United States was not abandoning space, just the shuttle. American astronauts would continue to fly to and from the International Space Station, but they would fly on the Russian Soyuz until the United States had a new spacecraft in service.

Over the years, the National Air and Space Museum has been called (approvingly and disapprovingly) a temple of technology, a testament to progress, and NASA's national visitors' center. It is difficult for the museum to venture far from those expectations. There was never any doubt that it would display *Discovery* as a remarkable flying machine and the space shuttle era as a time of lofty ambitions and accomplishments. Indeed, the museum chose to present two narratives: at the suburban Virginia location, *Discovery*'s "biography" as the champion of the shuttle fleet based on its record as the longest serving orbiter; and in the city, the shuttle's place in the trajectory of U.S. human spaceflight since the 1960s. Yet the curators also knew that it was necessary to address the two shuttle tragedies and the shuttle's most obvious shortcomings. The question was how best to do that to ensure that the shuttle story presented some of the texture of its history informed by scholarship.[16]

Well before *Discovery* arrived, the museum began developing a new gallery-size exhibition (4,500 square feet), funded primarily by NASA and called *Moving Beyond Earth,* in the building on the National Mall in Washington. There the spaceflight narrative flowed from very early concepts of a spaceplane and the origins of the shuttle in post-Apollo planning through its design, 135 missions for living and working in space, and the International Space Station to an updatable review of future possibilities in commercial spaceflight, robotic astronauts, and space tourism. The main theme conveyed that the space shuttle represented a new idea of spaceflight based on routine access to orbit for practical work. Parts of the exhibit dealt with doing scientific research in space, servicing the Hubble Space Telescope, and building the space station. A replica of the middeck crew cabin stocked with flown items gave visitors a sense of the living and working environment on the shuttle.

The exhibit team dedicated two areas to matching treatments of the *Challenger* and *Columbia* tragedies, each with a factual description of what happened, the causes and consequences, and special memorabilia from NASA and a member of the crew families. Those displays were not monumental, nor were they meant to be shrines; they were places to acknowledge that spaceflight includes risk, mistakes, and occasionally tragic losses of human life. Visitors encounter them, or not, as they make their way through the gallery; the display cases are not flagged as special attractions but treated with dignified restraint in context with the rest of the narrative.

Elsewhere, the team found ways to show that nothing about the shuttle was inevitable; that it reflected technical, political, and operational choices; and that in the final analysis it accomplished much but not all of what it was meant for. For example, a large display of concept models presented the evolution of the space shuttle through design and cost tradeoffs from a fully reusable to a compromised, partially reusable vehicle. Space station models made a similar point. The exhibit team sought to engage visitors in both the conceptual and operational history of shuttle-era spaceflight through problem-solving and decision-making interactives. Acting as a flight director, visitors tried to save a troubled mission while getting sometimes contrary advice from the mission control team, or tried to build and equip a space station module to meet its purpose yet stay within budget. While the general ambience and visuals of the exhibit gallery gave an impression of being in space, the material objects and interactives carried the narrative. The *Discovery* display at the museum's Udvar-Hazy Center and the exhibition on the Mall were complementary—one presenting the icon directly and the other interpreting its meaning.[17]

Situating the shuttle in a continuum from early spaceplane concepts before and during the Mercury-Gemini-Apollo era to the completion of the International Space Station, curators reached the conclusion that it was a long experiment in an alternative mode of spaceflight. Even in the NASA community, many had come to view the space shuttle as an experimental spacecraft. Having run that course, NASA by 2011 reverted to the earlier mode of leaving earth orbit in capsule-style spacecraft. With uncertainty about final vehicle design and flight plans, the gallery's forecast for post-shuttle human spaceflight had to be subject to updating to keep the exhibition current.[18]

The National Air and Space Museum presented the shuttle era as a period of practical living and working in space, not a time of exploration as traditionally conceived. The curators did not establish a romantic or idealized conceptual frame of meaning, such as pioneering the space frontier. Instead, they presented a generally positive factual narrative but acknowledged the reality that the shuttle did not meet all expectations and suffered two tragedies. They avoided the excesses of triumphalism, exceptionalism, and inevitable progress for which spaceflight advocates are often criticized. They took as their educational mission to present some insight into the challenges and choices in human spaceflight, the technical and political decisions that affected spacecraft design and flights, the changing role and look of the astronaut corps, the team of thousands that made the enterprise possible, and even the popular culture artifacts that appeared during the era. That historical approach was embedded in an interactive, immersive environment, for ultimately the public "owns" human spaceflight and should understand it. In the end, the museum's team believed they had interpreted the shuttle appropriately to their mission to "Commemorate, Educate, Inspire."

The other recipient museums also staged arrival events with dignitaries, astronauts, music, and flourishes customized to their locales, and they developed unique curatorial plans and exhibitions appropriate to their missions. After a flyover past the Statue of Liberty and Manhattan landmarks, the Intrepid Sea, Air & Space Museum received *Enterprise* by barge transport from John F. Kennedy Airport through Long Island Sound and up the Hudson River with maritime fanfare and a flotilla of escorts. *Enterprise* was lifted and perched in an extraordinary location, on the flight deck of the aircraft carrier *Intrepid,* where the collection and exhibits emphasize military aviation. The museum erected a large protective tent around it until a new space pavilion was readied. Until receiving *Enterprise,* the Intrepid Museum's primary link to spaceflight was the carrier's history as a recovery ship for two early space missions, Mercury-Atlas 7 and Gemini 3. The new exhibits around *Enterprise* addressed both the space shuttle and experimental flight testing, in recognition of the role *Enterprise* played in the shuttle program. Elevated enough to accommodate visitors and exhibits underneath, with a platform for close viewing at the nose, the prototype orbiter *Enterprise* became the focal point for narratives of engineering innovation. The museum's curatorial team distilled this meaning for the shuttle era from the orbiter that first embodied it. Extending that meaning into its educational mission and outreach to student visitors, the Intrepid Museum drew shuttle astronauts into its orbit for enhanced science, technology, engineering, and math (STEM) programs anchored by *Enterprise.* It also incorporated the engineering theme into its treatment of the two shuttle tragedies by explaining the role of analysis and testing in determining their causes.

The California Science Center received *Endeavour* by road trip from Los Angeles International Airport on a two-day parade through residential and business neighborhoods where people lined the streets to admire the shuttle. The Science Center placed the orbiter on its wheels inside a temporary building while planning and constructing a new building, slated to open in 2018, that would accommodate a vertical display. Having acquired two solid rocket boosters and the last real external tank, this exhibit team aimed to display *Endeavour* as a shuttle is most commonly remembered: on the launch pad or ascending toward space. The towering shuttle stack that signifies the power and magnitude of the space transportation system no doubt will create a sense of awe and also prime visitors for the related science and technology exhibits. According to this ambitious plan, viewers will be able to encounter the vehicle at various levels and peer into the partially open payload bay and the flight deck. The permanent exhibits will acknowledge the local history of the shuttle but primarily will be hands-on science-learning activities in accordance with this center's mission to "stimulate curiosity and inspire science learning in everyone." *Endeavour* will be the centerpiece for space-related science and technology education, including an aerospace learning lab. Its meaning in this setting is to be an educational beacon "to inspire a new generation of scientists, innovators, and explorers." For the California Science Center, acquisition of the space shuttle had a transformative effect on both its facility and its educational programming.[19]

Atlantis was towed with fanfare from the work area of Kennedy Space Center to the theme-park-like Visitor Complex just outside NASA's gate. There workers moved it into a huge new space exhibit building architecturally designed to suggest ascent, reentry, and the surface finishes of the space shuttle. Full-size replicas of the huge external tank and solid rocket boosters stood at the entry to the *Atlantis* building; inside, the orbiter "flew" high above at a forty-three-degree flight angle with its payload bay doors open as it appeared in orbit. In a dramatic reveal, visitors encountered this spectacular display after viewing an in-theater video orientation that immersed them in the sound and fury of a launch. The show ended with *Atlantis* in orbit, and as the curtain wall opened into the exhibition hall, there—surprise!—was the actual *Atlantis* almost close enough to touch. Visitors then wandered among exhibits related to its operations, including a stand-upon launch pad, a sit-inside flight deck and crawl-through space station module, and a glide-sloped slide to the ground floor to simulate the orbiter's descent to landing. There they could continue into the *Shuttle Launch Experience* attraction, a simulator programmed for countdown, ignition, and ascent into orbit that was thrilling in its realism.

Everything about the presentation there was "calculated to wow" visitors, to immerse them in realistic experiences along a dramatic, emotional arc. The Kennedy Space Center Visitor Complex shares the central Florida tourism market with Disney World, Universal Studios, and other theme park giants, and its presentations are inflected by that influence. The corporation that operates the Visitor Complex for NASA is a global presence in the hospitality management industry, also operating casinos, parks and resorts, stadiums, and other entertainment venues. While NASA staff provide the content and guidance, the contractor supplies creative talent and the latest presentation techniques to energize the content into thrilling "out-of-this-world" experiences. The *Atlantis* and *Shuttle Launch Experience* teams also had ready access to NASA's astronauts, engineers, and technicians as advisers to ensure a high degree of accuracy in exhibits and activities designed to be fun. This was by far the most elaborate, and most expensive, of the four orbiter displays.[20]

Everything here also related directly to the mission of the Kennedy Space Center. This exhibition uniquely focused on the resident workforce who serviced and launched the shuttle for three decades. These workers were quoted and pictured throughout the exhibits; they narrated videos and served as tour guides. Their presence yielded a very personal and people-centered presentation of the meaning of the shuttle and spaceflight in their lives and their community. This approach also accented the emotional experience of spaceflight as heard in their voices, seen on their faces, and directly communicated to the visitors. No other shuttle host museum could tell their stories with such authenticity and authority.

The most emotional experience occurred in the *Forever Remembered* hall that opened in 2015, a more daring astronaut memorial than any other. Collaborating closely with the families of the *Challenger* and *Columbia* crews, NASA's exhibit team designed a tribute to the individuals lost in those tragedies. The quiet, dimly lit sanctuary off the main hall featured fourteen display cases, each filled by the families with objects chosen to represent their astronaut's interests while alive. Unlike names carved in stone, these exhibits conveyed memories of vibrant individuals. The cases were mounted like portraits on the walls of an aisle that glowed with an ethereal blue light. At the end of the aisle, in an inner sanctum, one large window framed in the same blue light revealed a piece of *Challenger*'s fuselage with its painted American flag intact, and another light-framed window revealed *Columbia*'s bent and charred window frame (plate 17). These memorials to the destroyed shuttle vehicles caught visitors by surprise; this was the first place that debris from the accidents appeared on public display. The two large pieces appeared to be sculptures suspended weightless in space. They were not displayed to be prurient or macabre but to respect the vehicles that had "lived" next door at NASA; people generally viewed them in silence. Visitors departed through another aisle that told the story of recovery and return to flight, moving from tragedy to optimism about future spaceflight. Creating this vivid memorial hall was a brave and delicate labor for all involved, but one that many believed necessary for NASA and the public to accept the lessons of those losses. Not far from this building, the monumental outdoor space mirror that is the designated national memorial for astronauts is abstract and formal. This interior hall is intimate and emotional, a remembrance of lives and deaths, and ultimately a hopeful affirmation of human spaceflight.[21]

Four space shuttle orbiters in four locations, four different presentations, and countless memories filtered by those who curate and those who visit: what does it all mean? In each instance, the space shuttle was presented as an icon of national achievement, an engineering marvel and spaceflight technology unlike any other, and one that defined an entire generation of human activity in space. That was the principal meaning presented as public memory of this "magnificent flying machine."[22] From that point, the four institutions departed on their own courses—one toward engineering innovation, one toward science learning, another toward emotion and personal experience, and the Smithsonian toward a measured historical assessment leavened with scholarly judgment. Not one presented the complete or final meaning of human spaceflight in the shuttle era, but a visit to all four would yield a thoughtful understanding of the technology, its uses, and its place in history.

The end of the space shuttle program and the retirement of the orbiters to museums prompted a variety of other deliberate efforts to capture and communicate the recent legacy of human spaceflight. Some of these arose within the space shuttle technical community as concerted efforts to record memories through oral history and "knowledge capture" projects. Understandably, the first comers were often nostalgic and celebratory, paying uncritical tribute to what by any measure was a remarkable technical and operational achievement sustained for three decades. Other efforts arose outside the shuttle community among scholars training a historian's or political scientist's analytical eyes on the era.

A prime instance of internal memory-making resulted in a work of graphic art. As a coda to the shuttle era, a final space shuttle program icon commemorated the thirty years of flight history as an embroidered patch, lapel pin, decal, and decorative motif for a variety of souvenirs. At the center of the design a shuttle launch rose against a circle shaded to resemble the limb of the earth as viewed from orbit. Behind the launch, five narrow panels fanned out like the booster rockets that fell away from the vehicle stack during ascent. Seven stars scattered across two panels to the left and seven more to the right symbolized the fourteen spacefarers lost in the two shuttle tragedies. Five larger stars on one panel symbolized the five orbiters in the fleet. The design distilled the essence of what was most visibly memorable about spaceflight in the shuttle era—the awesome launches; the nation's winged orbiters, unlike anything else flown in space; and more subtly, the losses. The patch embodied thirty years of memory in a visual icon packed with meaning.

On a lighter note, NASA commissioned from popular comic strip artist Brian Basset a cartoon to commemorate the end of the space shuttle program. His "Red and Rover" strip often featured a boy and his dog flying in a cardboard

Fig. 7.4. The sheer joy and inspiration of spaceflight are captured in this cartoon commissioned by NASA to mark the end of the space shuttle program in 2012. NASA.

box on imagined space missions. Basset's cartoon "What a Ride it's Been!" featured the two playing with a shuttle orbiter, clearly exhilarated (fig. 7.4). These two celebratory mementos punctuated the end of the shuttle era with a message that it had been a great achievement.

During the period of transition and retirement, almost everyone inside NASA who was a veteran of the shuttle era became a self-styled historian, telling the stories and gathering the mementos of personal experience, and also reflecting on the broader significance to the nation of the long enterprise in "routine" spaceflight. The NASA centers launched formal efforts in "knowledge capture" interviews to record at least some of the lessons-learned experience of the shuttle-era workforce. Much of this material now resides in the respective archives where they were collected. In addition to the formal memory recording projects, Wayne Hale, a longtime NASA engineer, flight director, and manager who led the shuttle program's return to flight after the *Columbia* tragedy, spearheaded a volunteer effort to collect and publish personal recollections for a public volume. During his tenure as head of the space shuttle program, he requested that active and retired workers record many of the behind-the-scenes accomplishments that could help answer the question, "Was it worthwhile?"[23]

In addition to the two tragedies and the shuttle's inability to meet the primary expectations for routine spaceflight—lower cost and greater frequency of flights—another shadow hung over the shuttle-era workforce. It was bad enough for morale that longtime critics had railed against the shuttle and space station as mistaken endeavors. It was more demoralizing when NASA administrator Michael Griffin made such claims during his first year on the job in 2005 and thereafter. Then, what many viewed as the premature end of the space shuttle program galvanized some within the agency to go on record with a rebuttal. Hale and other shuttle program managers admitted that with hindsight, the shuttle's flaws were evident and routine spaceflight posed greater challenges than expected, but they determined that those caveats should not diminish "the most remarkable achievement of its time."[24]

Several hundred people contributed to *Wings in Orbit,* the book Hale encouraged, on their own time as "a labor of love," recording how they did many of the countless tasks involved in designing, building, and operating a fleet of reusable spacecraft; what they achieved and didn't; and from their perspective, what the shuttle contributed to technology, science, and society. As the insiders' history, it exudes pride and passion but also forthright explanations of how they met formidable engineering and operational challenges. The book was meant to be primary source material for others who may render judgment on the shuttle era. Hale concluded that a national space policy reconsideration of the nature of spaceflight that the shuttle enabled should have occurred after about ten years of flight. "The shuttle is what it is," he said, not what it should have been. In retrospect, it was an extended experiment on the edge of what was possible. A second-generation redesign might have resulted in a safer, more economical vehicle better suited to its intended purpose. While acknowledging NASA's flaws and errors, he also cited political leadership's role in the shuttle's shortcomings, but on balance he was satisfied that the legacy of the shuttle era was "pretty amazing."[25]

Some historians and scientists did not wait until the end of the shuttle era to begin taking the measure of its significance. From the beginning, some had been skeptical of the value of human spaceflight in earth orbit and had issued recurrent critiques of the shortcomings of the shuttle and space station programs. Some of the most strident critics—the respected physicist James A. Van Allen, for one—were themselves involved in space exploration but were proponents of robotic rather than human activity. They fiercely objected to the added cost and complexity of human spaceflight, symbolized by the shuttle, because it "bled dry" more scientifically productive space exploration projects.[26]

Van Allen raised uncomfortable questions that he thought worthy of a calm national debate: "Does human spaceflight continue to serve a compelling cultural purpose or our national interest? Or does it simply have a life of its own, without a realistic objective that is commensurate with its costs? Or, indeed, is human spaceflight obsolete?" He and others argued that human spaceflight was a romantic "ideology of adventure" without a practical rationale. Nor was it a productive return on investment in resultant discovery, knowledge, or benefits. Some provocatively ridiculed human spaceflight as a "stunt," "spectacle," or "nonsense."[27]

It is perhaps unfortunate that such critics were never appointed to a commission on the future of human spaceflight. Such commissions often were fairly homogeneous in outlook. Reasoned different perspectives from the periphery might have stimulated the framing of a more compelling rationale for a permanent human presence in space. Except for persistent questioning of the need for human spaceflight in interview soundbites, published opinion pieces and articles, and occasional lectures or congressional testimony, these critics were not able to galvanize a transformative national space policy debate. While their adversarial voices were not welcome in the dominant community that framed the meaning of spaceflight, their challenges to prevalent ideas are part of the permanent record and thus factor into the meaning of the shuttle era.

During the retirement year 2011, outside NASA a flood of tributes issued from all the expected places: the news media, the aerospace press, television networks, documentary producers, and book publishers. Almost without exception, these celebrated the shuttle era as a legacy of national achievement in space, with emphasis on successes and "firsts" and generally more attention to the shuttle as an engineering marvel than to the social impact of spaceflight. Examples in these genres include the special commemorative volume issued by *Air & Space* magazine, shuttle documentaries produced by the Discovery and Smithsonian television networks, and beautifully illustrated "coffee-table" books. Imagery of the shuttle is amenable to interpretations of grandeur; a launch is indeed an awesome spectacle even in still photos, the orbiter against the earth is a wondrous sight, and even landing conveys a sense of power, so these highly visual retrospectives tended toward a glowing triumphalist interpretation.[28]

Still to come are more analytical reflections on the meaning of spaceflight in the shuttle era, not by historians only, but one hopes by a cohort of scholars and interpreters in the arts, humanities, and social sciences. If prolonged academic attention to the space-race era is an indication, scholarly assessment of the shuttle era, already under way, may well continue for decades, too. One such project undertaken by the American Institute of Aeronautics and Astronautics tapped historians, social scientists, and engineers to assess the space shuttle legacies and lessons learned. The authors were a mix of insiders and outsiders who offered somewhat more critical answers to questions about the achievements and meaning of spaceflight in the shuttle era. Situating the shuttle and spaceflight into broader contexts shows how memory and meaning are dynamic; both are influenced by the lenses through which the historical moment is viewed.[29]

It is not too early to speculate what other assessments may be, for some have percolated through the shuttle era. The idea of routine spaceflight will come under scrutiny, as it has since the beginning, more keenly since the *Challenger* accident and ensuing investigation, and with the further record of twenty-five more years of shuttle flight experience, a second calamitous loss, and another return to flight. What was not publicly expressed before 1986 is now commonly held: that the shuttle was inherently an experimental vehicle that operated on the margins of what was technically achievable. By its nature it could never be flown often enough and tested thoroughly enough to reach the confidence level for truly routine, airline-like service. The suspension of the spaceflight participant program after 1986 codified that admission; thereafter, the agency permitted no one but professional astronauts to fly. Finally, even astronauts and top managers publicly admitted that the shuttle was always experimental.

Routine spaceflight proved to be an illusion. Despite falling short of that ideal, however, spaceflight in the shuttle era achieved a pattern of more frequent and varied missions that suggested what routine spaceflight might be like in the future. If technical and operational lessons learned are factored into new programs or if new technologies resolve the issues that plagued the shuttle with costly delays, that goal may yet be achievable. The downside of routine spaceflight, though, is that the public generally loses interest in or has little awareness of its purpose, which presents a challenge for sustaining public support. Routine, affordable flights for space tourists, however, might someday overcome such diffidence.

The idea of reusability as the key to economy is also likely to be scrutinized from various perspectives. The great disappointment of the shuttle era was that the cost of spaceflight did not decrease. In theory, it should have, but in practice the shuttle was not the right kind of reusable spacecraft to drive down costs. It required too much servicing between missions, and the time needed to prepare it for flight was never consistent enough to permit a reliably regular launch schedule. Also, despite attentive servicing on the ground, the vehicle often suffered equipment problems in flight. There were always kinks to be worked out, almost as if it were a new vehicle each time rather than the same one in reuse. Determining whether the path taken toward the goal of routine spaceflight in an economical vehicle was worthwhile will no doubt engage analysts of the shuttle era.

Policy is another lens for viewing the shuttle era. Several scholars offered early assessments of the legacy of the space shuttle years before the program ended. Space policy analyst John Logsdon dissected the decision to proceed with development of a space shuttle and argued that it was a policy failure. He cited not only mistaken ideas about reusability and routine flight but also the government's poor commitment to this major technological endeavor through inadequate funding and shaky political support. Policy failure does not correspond to technical or operational failure, but it does mean that the shuttle concept was compromised from the start and thus its touted promise had little chance of being fulfilled. In that sense, spaceflight in the shuttle era was something of a fantasy, doomed not to be fully realized.[30]

The idea of commuting to space to do useful work and living there to do research also will be assessed as part of the shuttle-era legacy. Shuttle missions and International Space Station expeditions constitute the database for this assessment. From the perspective of multipurpose space activity, the spaceflight record is filled with successes. Astronauts carried out increasingly sophisticated work in space, culminating in the mechanical and electrical assembly of the huge space station and outfitting its interior. This effort was without precedent on its largest and smallest scales. The series of servicing missions to rejuvenate the Hubble Space Telescope undoubtedly will stand the test of time as a major legacy of shuttle-era spaceflight. Likewise, the retrieval, repair, and release or return of satellites and the deployment and recovery of free-flying research satellites (the SPARTANs, SPAS, and others) posed unique challenges and were executed well.

As historians and other analysts drill deeper into mission details, instances of exceeded and unmet expectations will add texture to the legacy of human spaceflight during the shuttle era. To date, the least examined and understood activity, outside the scientific communities involved, is research on the shuttle and space station. Until scholars from various fields examine the record of research productivity, the significance of these spacecraft as laboratories and observatories remains vague. Clearly they afforded many opportunities for scientists and engineers to pursue investigations in microgravity and the vacuum of space, but notable results of that research are not yet readily accessible and intelligible outside the involved disciplines. Even so, the expansion of scientific research well beyond that carried out during earlier U.S. and Soviet/Russian human spaceflight programs is in itself a positive accomplishment. From a broad perspective, the use of space as a laboratory environment is part of the shuttle-era legacy. It deserves to be assessed.

In its social impact, the shuttle era was transformational by opening spaceflight to women and Americans of African, Hispanic, Asian, and Native heritage as full participants. Their admission into the astronaut corps also helped to open up the rest of the spaceflight enterprise to full participation as technicians and engineers, crew trainers, flight controllers and mission support personnel, and managers in all areas and levels. The NASA and contractor workforce looked much different in 2011 than in 1981, partly due to broad social changes, but the shuttle program during that era certainly sparked the demographic transformation. The shuttle era also linked the U.S. space agency in partnerships with those of Russia, Europe, Canada, and Japan. It enabled other nations already active in exploratory space science to develop spacefaring capabilities, to build their own astronaut corps and shared spacecraft hardware, and to join the community of human spaceflight. Cooperation in the shuttle and space station programs brought human spaceflight into the realm of foreign policy and diplomacy more overtly than in the space-race years, and probably more permanently.[31]

There are essential paradoxes in the meaning of human spaceflight in the shuttle era. If indeed history becomes a myth kept alive and elaborated by its retelling, the heroic myth of pioneering the space frontier from the 1960s persisted into the history of the space shuttle era. Spaceflight in and beyond that "new era" continued to be an adventure myth. NASA and its various advisory committees have tried other metaphors of work, utility, practicality, even logic, but none proved resonant enough in the public psyche to dislodge the frontier myth, which remains the basic chord upon which themes and variations are played. The strongest alternative framework of meaning for human spaceflight appeared at the dawn of the shuttle era—the pragmatic vision of routine commuting to live and work in near-earth orbit, exploiting, not exploring, space for useful benefits. The shuttle as a space truck became the icon for that mindset, although the intended analog was an airline. The idea of routine spaceflight prevailed until the shocking *Challenger* tragedy exposed its naïveté. In sorrow, almost everyone from the president to the proverbial "man in the street" reverted to the familiar concept of pioneering the frontier in eulogizing the crew as brave explorers and acknowledging the inherent risk of spaceflight.

Commuting to deliver satellites and do similar yeomen's work no longer seemed an endeavor worth risking astronauts' lives compared with a more adventurous role as explorers.[32]

When spaceflight resumed after the loss of *Challenger,* the ideas of routine commuting and America's space truck service were set aside in favor of another pragmatic concept that was closer to the spirit of exploration—scientific research. The shuttle as a laboratory for scientific research became the dominant framework of meaning; it correlated with the majority of missions in the 1990s and also set the scene for "the next logical step"—a space station. As the space station suffered the winds of political fortune, shrinking in size and capability, the laboratory concept continued to fit human spaceflight well into the 2000s.

Various advisory committees tried other frameworks for human spaceflight—as the key to leadership, for example—but the frontier continued to be the favored "go-to" concept. Although among historians, the history of the frontier experience has become nuanced and problematic, it remains an idealized vision in the spaceflight community and arises as automatically as a reflex. As the founding generation of real space pioneers ages away and American society becomes increasingly diverse and untutored in nineteenth-century U.S. history, the frontier myth may gradually lose its power, but that has not happened yet. The question is whether NASA or the commercial spaceflight firms will move on to frame new meanings as (or if) human spaceflight continues in new modes. What concepts will fit space tourism? Will commercial spaceflight succeed in appropriating the rhetoric of routine spaceflight if it can establish regular service at a lower cost? Will the frontier gradually fade from spaceflight ideology as it has from this well-explored and populated planet? Answers to such questions may come soon.

Ultimately, the iconic status of the shuttle may prove to be as enigmatic as its shape that straddled the line between spacecraft and airplane. The space transportation system meant to operate like an airline; the reusable craft that would make spaceflight economical but didn't; the commuter van/delivery truck/service station/laboratory meant to make space travel routine but did not keep a regular schedule; the vehicle that was to be safe enough that "ordinary people" could fly along as passengers but took fewer than sixty non-astronauts on its 135 missions—these paradoxes tint the shuttle era, not as failures but as evidence of the difficulty of shaping meaning. To be effective, icons that symbolize grand ideas not only must resonate with widely held beliefs and values; they also must track with reality. In the shuttle era, reality proved to be a much harder thing to control than the optimists who launched a new era in spaceflight ever anticipated.[33]

And so it falls to scholars, museum professionals, and educators primarily to preserve the history of the space shuttle era and present its meaning to the public. Increasing temporal distance from the lived experience of the shuttle era will foster increased critical distance in interpreting its meaning. As the population shifts youthward, the stewards of shuttle and spaceflight history may have to find new framing rhetoric to engage those with no direct memory. These stewards must stay alert to the meaning of spaceflight as it evolves in the near future. As the United States population balance shifts, there are many Americans by birth and by choice who may not be acculturated in America's traditional myths and symbols; they may not intuitively understand the frontier metaphor or why spaceflight matters in the present and future. Will they concur that human spaceflight is in the nation's interest and the United States should continue to lead, or will they have more pressing national priorities? Will coming generations participate in, and lead, this ongoing American tradition or step back? Will they frame new rationales for human spaceflight? If global climate change or epidemic disease or terrorism threatens humanity's tenure on earth, might their motivating rationale for spaceflight be survival off the planet?

As memory institutions—the primary keepers and presenters of public memory—museums cannot predict what the future holds, but they can stay attuned to present trends and contexts related to their collections and exhibitions. They can continue to shed light on the past through active research and updated interpretations based on new knowledge. They are obliged to respect the dynamic nature of memory and meaning and to engage their audiences thoughtfully, even provocatively. The four museums that now display space shuttle orbiters as national icons and present varied narratives of human spaceflight especially bear this responsibility.

Notes

INTRODUCTION

1. Roger D. Launius, "Public Opinion Polls and Perceptions of U.S. Human Spaceflight," *Space Policy* 19 (2003): 163–75.
2. Revolutionary works that became influential across disciplines include Peter L. Berger and Thomas Luckmann, *The Social Construction of Reality: A Treatise in the Sociology of Knowledge* (London: Penguin, 1967), and Thomas S. Kuhn, *The Structure of Scientific Revolutions* (Chicago: University of Chicago Press, 1962), which is still popular in a fiftieth-anniversary edition. More recent works, such as Stephen Bann, *The Inventions of History* (Manchester: Manchester University Press, 1990), and John R. Searle, *The Construction of Social Reality* (New York: Free Press, 1995), also make the case that knowledge, perception, meaning, and even what is assumed to be factual are fluid, not absolute.
3. Much of this material is readily accessible in the NASA Historical Reference Collection at NASA Headquarters in Washington, D.C.; in the Johnson Space Center History Collection at the University of Houston's Clear Lake campus in Texas; and online through the NASA History website, www.history.nasa.gov. The U.S. Human Spaceflight History portal (www.jsc.nasa.gov/history/hsf_history.htm) links to many compilations of factual information and chronologies. Shuttle-era materials housed at the National Archives and Records Administration Federal Records Center in Suitland, Maryland, and the National Archives site in College Park, Maryland, can be accessed, but less conveniently because many are still being processed and catalogued.

4. Media studies within the fields of political science and communications yield both theoretical concepts and quantitative methods of discourse analysis. Among the scholars in media studies are James W. Carey, Robert Entman, William Gamson, and Eric Louw, who illuminate how elite discourse, issue framing, and agenda setting influence the construction of meaning by the news media.
5. Robert Benford, Erving Goffman, John Searle, David Snow, and others have opened windows into the ways that organizations frame, spread, and tend to their interests by using rhetorical strategies to achieve their goals. Sonja Foss and John Louis Lucaites are among those who have energized the study of rhetoric. Historians, political scientists, rhetoricians, and speechwriters attend to the crafting and influence of presidential rhetoric, as in cited works on the shuttle-era presidents. Martin Foss, George Lakoff, Andrew Ortony, Paul Ricoeur, and Susan Sontag examine the role of metaphor in public life. The works of Richard Dawkins and Susan Blackmore are suggestive for thinking about memes of human spaceflight.
6. Cornelius Castoriadis, George Marcus, John R. Searle, Claudia Strauss, and Charles Taylor have explicated the idea and role of imaginaries, a concept that has spread from philosophy into a variety of disciplines in contemporary scholarship.
7. Malcolm Barnard, Roland Barthes, Nicholas Mirzoeff, Marita Sturken, and others have contributed to the growth of visual culture studies that interpret images (art, photography, television, film, and other sources) as significant cultural texts.

CHAPTER 1. SPACEFLIGHT

1. The heroic age in space is a premise in Martin J. Collins, ed., *After Sputnik: 50 Years of the Space Age* (New York: Smithsonian Books/HarperCollins, 2007), and Michael J. Neufeld, ed., *Spacefarers: Images of Astronauts and Cosmonauts in the Heroic Era of Spaceflight* (Washington, DC: Smithsonian Institution Scholarly Press, 2013). It is the "Golden Age" in Frederick I. Ordway III and Randy Liebermann, eds., *Blueprint for Space: Science Fiction to Science Fact* (Washington, DC: Smithsonian Institution Press, 1992).
2. Benedict R. Anderson, *Imagined Communities: Reflections on the Origin and Spread of Nationalism* (New York: Verso, 2006); Cornelius Castoriadis, *The Imaginary Institution of Society*, trans. Kathleen Blamey (Cambridge, MA: MIT Press, 1987); George E. Marcus, *Technoscientific Imaginaries* (Chicago: University of Chicago Press, 1995); Dilip Parameshwar Gaonkar, "Toward New Imaginaries: An Introduction," *Public Culture* 14/1 (2002): 1–19; Charles Taylor, *Modern Social Imaginaries* (Durham, NC: Duke University Press, 2004); Claudia Strauss, "The Imaginary," *Anthropological Theory* 6/3 (2006): 322–44.
3. Roger D. Launius, "Escaping Earth: Human Spaceflight as Religion," *Astropolitics* 11/1–2 (2013): 45–64; Kendrick Oliver, *To Touch the Face of God: The Sacred, the Profane, and the American Space Program, 1957–1975* (Baltimore: Johns Hopkins University Press, 2013). Such exemplary studies of American cultural myths and metaphors (or imaginaries) as Henry Nash Smith, *Virgin Land: The American West as Symbol and Myth* (Cambridge, MA: Harvard University Press, 1950), and Leo Marx, *The Machine in the Garden: Technology and the Pastoral Ideal in America* (New York: Oxford University Press, 1964), remain suggestive for approaching spaceflight and other topics, as does Martin Foss,

Symbol and Metaphor in Human Experience (Princeton, NJ: Princeton University Press, 1949).

4. Classic works include Peter L. Berger and Thomas Luckmann, *The Social Construction of Reality: A Treatise in the Sociology of Knowledge* (London: Penguin, 1967); John R. Searle, *The Construction of Social Reality* (New York: Free Press, 1995); Erving Goffman, *Frame Analysis: An Essay on the Organization of Experience* (New York: Harper & Row, 1974); Robert D. Benford and David A. Snow, "Framing Processes and Social Movements: An Overview and Assessment," *Annual Review of Sociology* 26 (2000): 611–39.
5. Among the best in the technical genre are Dennis R. Jenkins, *Space Shuttle: The History of the National Space Transportation System: The First 100 Missions* (North Branch, MN: Specialty Press, 2001), and T. A. Heppenheimer, *History of the Space Shuttle* (Washington, DC: Smithsonian Institution Press, 2002). In the popular genre, Piers Bizony, *The Space Shuttle: Celebrating Thirty Years of NASA's First Space Plane* (Minneapolis: Zenith Press, 2011), and David M. Harland, *The Space Shuttle: Roles, Missions and Accomplishments* (Chichester, UK: John Wiley & Sons, 1998), are commendable. Mark E. Byrnes, *Politics and Space: Image Making by NASA* (Westport, CT: Praeger, 1994); James L. Kauffman, *Selling Outer Space: Kennedy, the Media, and Funding for Project Apollo, 1961–1963* (Tuscaloosa: University of Alabama Press, 1994); David Meerman Scott and Richard Jurek, *Marketing the Moon: The Selling of the Apollo Lunar Program* (Cambridge, MA: MIT Press, 2014), examine NASA's methods of persuasion.
6. Thomas Rosteck, ed., *At the Intersection: Cultural Studies and Rhetorical Studies* (New York: Guilford Press, 1999); Deborah Tannen, ed., *Framing in Discourse* (New York: Oxford University Press, 1998); Zhongdang Pan and Gerald M. Kosicki, "Framing Analysis: An Approach to News Discourse," *Political Communication* 10/1 (1993): 55–75; William A. Gamson et al., "Media Images and the Social Construction of Reality," *Annual Review of Sociology* 18 (1992): 373–93; William A. Gamson and Andre Modigliani, "Media Discourse and Public Opinion on Nuclear Power," *American Journal of Sociology* 95 (1989): 1–37; C. Whan Park, Bernard J. Jaworski, and Deborah J. MacInnis, "Strategic Brand Concept-Image Management," *Journal of Marketing* 50 (1986): 135–45; Andrew Ortony, *Metaphor and Thought,* 2nd ed. (Cambridge: Cambridge University Press, 1993); George Lakoff and Mark Johnson, *Metaphors We Live By* (Chicago: University of Chicago Press, 1980); Paul Ricoeur, *The Rule of Metaphor: Multi-Disciplinary Studies of the Creation of Meaning in Language,* trans. Robert Czerny (Toronto: University of Toronto Press, 1975); Maurice Halbwachs, *On Collective Memory* (Chicago: University of Chicago Press, 1994); Michael Kammen, *Mystic Chords of Memory: The Transformation of Tradition in American Culture* (New York: Alfred A. Knopf, 1991); Malcolm Barnard, *Approaches to Understanding Visual Culture* (New York: Palgrave, 2001); Roland Barthes, "The Rhetoric of the Image," in *Image, Music, Text,* trans. Stephen Heath (New York: Hill and Wang, 1977), 32–51.
7. Thoughtful discussions of the nature of myth are found in James Oliver Robertson, *American Myth, American Reality* (New York: Hill and Wang, 1980); William H. McNeill, "The Care and Repair of Public Myth," *Foreign Affairs* 61/1 (1982): 1–13, also in Nicholas Cords and Patrick Gerster, eds., *Myth and the American Experience* (New York: HarperCollins, 1991), 435–45; "Myth and Historical Memory," in Richard Slotkin, *The Fatal Environment: The Myth of the Frontier in the Age of Industrialization, 1800–1890* (New York: Atheneum, 1985), 13–32; William H. McNeill, "Mythistory, or Truth, Myth, History, and Historians,"

American Historical Review 91 (1986): 1–10; Eric Hobsbawm and Terence Ranger, eds., *The Invention of Tradition* (New York: Cambridge University Press, 1992); and Peter Heehs, "Myth, History, and Theory," *History and Theory* 30 (1994): 1–19.

8. Willy Ley, *The Conquest of Space* (New York: Viking Press, 1949); Annual Symposia on Space Travel held at the American Museum of Natural History, Hayden Planetarium, in New York City, October 1951 and October 1952; Cornelius Ryan, ed., *Across the Space Frontier* (New York: Viking Press, 1952) and *Conquest of the Moon* (New York: Viking Press, 1953). *Collier's* magazines featuring space travel were issued in 1952 (March 22 on conquest of space, October 18 and 25 on conquering the moon), 1953 (February 28, March 7, and March 14 on astronauts and space travel, and June 27 on a small orbital research station), and 1954 (April 30 on exploration of Mars). Disney offered Sunday evening television broadcasts of three films inspired by the *Collier's* articles: "Man in Space" (March 9, 1955), "Man and the Moon" (December 28, 1955), and "Mars and Beyond" (December 4, 1957). Randy Liebermann, "The *Collier's* and Disney Series," in Ordway and Liebermann, *Blueprint for Space,* 135–46; Howard McCurdy, *Space and the American Imagination* (Washington, DC: Smithsonian Institution Press, 1997). "The Conquest of Space: A Conversation Between Wernher von Braun and Willy Ley," recorded in 1959; Vox Sound Recording DL522 (1962), accessed at archive.org/details/AConversation BetweenDr.WernherVonBraunAndWillyLey.
9. Ryan, *Across the Space Frontier,* xi–xiv, 44–45, 50–56; Michael J. Neufeld, *Von Braun: Dreamer of Space, Engineer of War* (New York: Alfred A. Knopf, 2007), especially Chapter 11, "Space Superiority," 246–78. The founder of the American Interplanetary Society linked conquest and space in a book title almost twenty years earlier, but he did not use the term in a military sense: David Lasser, *The Conquest of Space* (New York: Penguin Press, 1931). The National Geographic Society later published a book with a similar title: William R. Shelton, *Man's Conquest of Space* (Washington, DC: National Geographic Society, 1968). The phrase "conquest of space" is still in parlance, although usually without a Cold War aura.
10. Wernher von Braun, "Prelude to Space Travel," in *Across the Space Frontier,* ed. Cornelius Ryan (New York: Viking Press, 1952), 12–70; and Willy Ley, "A Station in Space," in Ryan, *Across the Space Frontier,* 98–117 (quotes from 100 and 117); "What Are We Waiting For?," editorial preface in the *Man Will Conquer Space Soon* issue of *Collier's* (March 22, 1952).
11. McCurdy, *Space and the American Imagination;* Wernher von Braun, "Crossing the Last Frontier," *Collier's,* March 22, 1952, pp. 22–23, 72–73; John M. Logsdon, ed., *Exploring the Unknown: Selected Documents in the History of the U.S. Civil Space Program,* vol. 1: *Organizing for Exploration* (Washington, DC: National Aeronautics and Space Administration, 1995), 176; Liebermann, "The *Collier's* and Disney Series," 135–46.
12. Dwayne A. Day, "Paradigm Lost," *Space Policy* 11/3 (1995): 153–59; Roger D. Launius and Howard E. McCurdy, *Imagining Space: Achievements, Predictions, Possibilities, 1950–2050* (San Francisco: Chronicle Books, 2001), 26–33.
13. David Farber, *The Age of Great Dreams: America in the 1960s* (New York: Hill and Wang, 1994); David Farber, ed., *The Sixties: From Memory to History* (Chapel Hill: University of North Carolina Press, 1994); David Farber and Beth Bailey, eds., *The Columbia Guide to America in the 1960s* (New York: Columbia University Press, 2001).

14. William E. Burrows, *This New Ocean: The Story of the First Space Age* (New York: Random House, 1998), and Walter A. McDougall, *The Heavens and the Earth: A Political History of the Space Age* (New York: Basic Books, 1985).
15. Frederick Jackson Turner, "The Significance of the Frontier in American History," American Historical Association address (1893) reproduced in Turner, *The Frontier in American History* (New York: Holt Rinehart Winston, 1920), 1–38. Allan G. Bogue, "Frederick Jackson Turner Reconsidered," *History Teacher* 27 (1994): 195–231, offers a chronology with commentary on scholarship and the Turner thesis.
16. William H. Goetzmann, "Exploration's Nation: The Role of Discovery in American History," in Daniel J. Boorstin, ed., *American Civilization* (New York: McGraw Hill, 1972), 11–36.
17. Howard E. McCurdy and Roger D. Launius, *Spaceflight and the Myth of Presidential Leadership* (Urbana: University of Illinois Press, 1997); Linda T. Krug, *Presidential Perspectives on Space Exploration: Guiding Metaphors from Eisenhower to Bush* (New York: Praeger, 1991).
18. His acceptance speech at the 1960 Democratic convention established the theme of the New Frontier. Theodore C. Sorensen, ed., *"Let the Word Go Forth": The Speeches, Statements, and Writings of John F. Kennedy* (New York: Delacorte Press, 1988), 96–102, especially 100–102. In *Kennedy: The Classic Biography* (New York: Harper & Row, 1965), 167, Sorensen comments that the New Frontier phrase and concepts were new to this speech and were Kennedy's own idea; they did not appear in prior campaign speeches or drafts by other writers. In the same book, he explains what space and the frontier meant to JFK (523–29). On Kennedy's use of the frontier for space, also see Kauffman, *Selling Outer Space,* and Krug, *Presidential Perspectives on Space Exploration.*
19. Example editorials and articles from the *New York Times* include: "The New Frontier," September 8, 1951, p. 12; "Radio Waves Lift Curtain on Space: Man's Frontiers Extended," December 28, 1953, p. 18; reviews of *Across the Space Frontier* by Cornelius Ryan et al., October 5, 1952, p. BR14, and *Frontiers of Astronomy* by Fred Hoyle, September 25, 1955, p. BR22; and from the *Washington Post,* "Man Now Exploring Frontiers of Space," January 7, 1951, p. B2. Editorial cartoons: "The New Frontier," *New York Times,* October 1, 1957; Fischetti, "U.S. Space Program," *Daily Times* (Melbourne, FL), October 13, 1959; Herblock, "They Went Thataway—New Frontier Space Program," *Washington Post,* December 30, 1960.
20. Excerpts from John F. Kennedy, address on "Urgent National Needs" to a joint session of Congress, May 25, 1961, reproduced in Logsdon, *Exploring the Unknown,* vol. 1, 453. Full text in *Public Papers of the Presidents of the United States, John F. Kennedy, 1961* (Washington, DC: Office of the Federal Register, National Archives and Records Administration, 1963).
21. John M. Logsdon, *The Decision to Go to the Moon: Project Apollo and the National Interest* (Cambridge, MA: MIT Press, 1969), and Logsdon, *John F. Kennedy and the Race to the Moon* (New York: Palgrave Macmillan, 2010). Sorensen, *"Let the Word Go Forth,"* 1. For detailed accounts of the preparation of Kennedy's speeches, see Theodore C. Sorensen, *Counselor: A Life at the Edge of History* (New York: HarperCollins, 2008), especially 223–31 on his rules of speechwriting and 359–372 on crafting the Inaugural Address; Thurston

Clarke, *Ask Not: The Inauguration of John F. Kennedy and the Speech That Changed America* (New York: Henry Holt and Company, 2004); Richard J. Tofel, *Sounding the Trumpet: The Making of John F. Kennedy's Inaugural Address* (Chicago: Ivan R. Dee Publisher, 2005).

22. Kennedy, "Urgent National Needs."
23. Ibid. and John F. Kennedy, address at Rice University in Houston on "The Nation's Space Effort," September 12, 1962, in *Public Papers of the Presidents of the United States, John F. Kennedy, 1962.*
24. Space-race editorials and headlines from the *Washington Post* before 1961 included: "The Guided Missile Race," August 18, 1954, p. 11; "Space Control Race," October 18, 1957, p. A18; "On Catching Up," November 11, 1957, p. A14; "Not a Defeat—a Challenge," November 16, 1957, p. A14; "Race for the Moon," February 21, 1958, p. A25; "U.S. Still Losing in Space Race," July 7, 1959, p. B15; "Catching Up," February 16, 1960, p. A14. Examples from the *New York Times* included: "Scientists Warn of Race to Moon," October 13, 1957, p. 32; "U.S. Found to Lag in Race to Space," November 6, 1957, p. 13; "Race into Outer Space: Where the U.S. and the Russians Stand," February 2, 1958, p. E5; "U.S. Lags in Space Race 8 Months After Sputnik I," May 25, 1958, p. 1; "Race into Space," December 30, 1958, p. 34; "Russian Again Ahead in Race into Space," January 4, 1959, p. E8; "Competing in Space," October 22, 1959, p. 36. "Space Exploration: U.S. v. Russia," *TIME,* January 19, 1959, cover. Editorial cartoons on the missile race included Herblock, "It's Just a Matter of Space," *Washington Post,* November 10, 1957, and "Can You Hurry? We're in an Important Race," *Washington Post,* January 29, 1957; Russell, "The Race Is to the Swift," *Los Angeles Times,* reprinted in *New York Times,* April 20, 1958; and Fischetti, "Who'll Be There First?" *New York Times,* October 12, 1958.
25. Examples of the space conquest theme from the *New York Times* included: "Man into Space," December 5, 1957, p. 34; "Conquests of Space Mapped by the U.S.," May 4, 1958, p. E7. Examples from the *Washington Post* included: "Conquest of Space," October 7, 1957, p. A8; "Space Control Race," October 18, 1957, p. A18; "Not a Defeat—A Challenge," November 26, 1957, p. A14; "Peril of the Hour," January 9, 1958, p. A16.
26. "U.S. Is Going All-Out to Win Space Race, Land on Moon in '67," *Washington Post,* May 26, 1961, p. 18; Herblock, "Fill 'Er Up—I'm in a Race," *Washington Post,* May 24, 1961.
27. "To the Moon and Beyond," *New York Times,* May 26, 1961, p. 32; "State of Union II," *New York Times,* May 28, 1961, p. E1.
28. "58% Oppose Moon Shot Proposal," *Washington Post,* June 2, 1961, p. A15.
29. "Stuntsmanship," *The Nation,* June 10, 1961.
30. John D. Morris, "First Kennedy Congress Responded Coolly to the Call of the New Frontier; Appeals to Blaze Trails Rebuffed," *New York Times,* September 28, 1961, p. 32.
31. For example, "A Call to Action," *New York Times,* October 18, 1957, p. 22; "Apathy About Space," *New York Times,* June 1, 1958, p. E8; Hanson W. Baldwin, "Neglected Factor in the Space Race," *New York Times Section Magazine,* January 17, 1960, pp. 15, 76–77.
32. "It will not be one man going to the moon . . . it will be the entire nation." Kennedy, "Urgent National Needs."
33. Brian Duff, "Storytellers of Space: NASA and the Media," in Martin J. Collins and Sylvia K. Kraemer, eds., *Space: Discovery and Exploration* (Washington, DC: Hugh Lauter

Levin Associates, 1993), 223–59; James Kauffman, "NASA's PR Campaign on Behalf of Manned Space Flight, 1961–63," *Public Relations Review* 17/1 (Spring 1991): 57–68.

34. Brian Duff, "Storytellers of Space: NASA and the Media," in Collins and Kraemer, eds., *Space: Discovery and Exploration,* 231–36.

35. Mercury Astronaut Group Portrait (variously date-coded as 61-MR4–7x, S62–08774, L-1989–00361, or EL-1997–00089). Contrast this image with the more conventional suit-and-tie group portrait on the cover of *LIFE* magazine, "The Astronauts," September 14, 1959.

36. Kauffman, Chapter 4, "*LIFE:* NASA's Mouthpiece in the Popular Media," Chapter 4 in *Selling Outer Space,* 68–92.

37. "Apollo 11 Launch," NASA photograph 69PC-0397, July 16, 1969; "Apollo 17 Night Launch," NASA photograph S72–55070, December 7, 1972.

38. "Race for the Moon" cover and "Poised for the Leap" cover story, *TIME,* December 6, 1968.

39. Buzz Aldrin on the Moon, NASA photograph AS11–40-5903, July 20, 1969; Buzz Aldrin and the U.S. Flag on the Moon, NASA photograph AS11–40-5875, July 20, 1969; both photographs taken by Neil Armstrong.

40. "Man on the Moon," *TIME,* July 25, 1969, cover.

41. "On the Moon," *LIFE,* August 8, 1969, cover, and "To the Moon and Back," *LIFE,* August 10, 1969, cover.

42. "Earthrise, Apollo 8," NASA photograph 68-HC-870, December 29, 1968. For an explication of this earth image and the next one, see Denis Cosgrove, "Contested Global Visions: *One-World, Whole-Earth,* and the Apollo Space Photographs," *Annals of the Association of American Geographers* 84/2 (1994): 270–94.

43. "Whole Earth," NASA photograph AS17–148-22727, December 7, 1972. Cosgrove, "Contested Global Visions."

44. Robert Hariman and John Louis Lucaites, *No Caption Needed: Iconic Photographs, Public Culture, and Liberal Democracy* (Chicago: University of Chicago Press, 2007); Guenther Kress and Theo van Leeuwen, *Reading Images: The Grammar of Visual Design,* 2nd ed. (London: Routledge, 2006); Nicholas Mirzoeff, *Introduction to Visual Culture,* 2nd ed. (London: Routledge, 2009).

45. Krug, *Presidential Perspectives on Space Exploration,* especially Chapter 2, "The Race to Space," 24–30, 42–43.

46. Republican senators Everett Dirksen (Illinois), Barry Goldwater (Arizona), and Jacob Javits (New York), plus Democratic senators J. W. Fulbright (Arkansas) and William Proxmire (Wisconsin), contested the space race as too costly. In 1963 the Senate Republican Policy Committee published a space critique titled "A Matter of Priority." "Congress Wary on Cost, But Likes Kennedy Goals," *New York Times,* May 26, 1961, p. 1; "Space Debate Grows Sharper: Budget Cuts Point Up Mounting Doubts About Race to the Moon," *New York Times,* October 6, 1962, p. 191; "Congress Has Second Thoughts on Space Funds," *New York Times,* April 7, 1963, p. E10; D. S. Greenberg, "Space Budget: Opposition Grows as Scientists, Congressmen, Voice Concern About Lunar Landing Goal," *Science* 140 (1963): 790–91; D. S. Greenberg, "Space: Formidable Political Base Overshadows Attempts to Revise Administration's Lunar Program," *Science* 145 (1964):

137–39; D.S. Greenberg, "After the Moon Landing: Senate Hearings Open Way for Debate," *Science* 150 (1965): 1003–5.

47. "Space and Serendipity," *New York Times,* May 24, 1961, p. 40; "The Lag in Space," *Washington Post,* August 24, 1962, p. A12.

48. Cartoons of this type appeared in various newspapers from 1965 on and especially in 1969, the year of the first lunar landing. The Apollo 17 cartoon by Crockett appeared in the Washington *Evening Star,* December 5, 1972, p. A12. The "one small step" cartoon by John Fischetti appeared in the *Chicago Daily News* on the same day, December 5, 1972.

49. "Man on the Moon," *Washington Post,* May 7, 1963, p. A16.

50. NASA–National Academy of Sciences Space Science Summer Study (1962) cited in D. S. Greenberg, "Space Budget: Opposition Grows as Scientists, Congressmen, Voice Concern About Lunar Landing Goal," *Science* 140 (1963): 790–91; quotes from D. S. Greenberg, "Space Program: Skepticism Grows but in Context of Cold War It Is Hard for Congress to Say No," *Science* 139 (1963): 890–91; D. S. Greenberg, "Space: Caution Prevails on Post-Apollo Commitments," *Science* 153 (1966): 1221–22.

51. Amitai Etzioni, *The Moon-Doggle: Domestic and International Implications of the Space Race* (Garden City, NY: Doubleday, 1964), ix.

52. "58% Oppose Moon Shot Proposal," *Washington Post,* June 2, 1961, p. A15; Herbert E. Krugman, "Public Attitudes Toward the Apollo Space Program, 1965–1975," *Journal of Communications* 27 (1977): 87–93; Roger D. Launius, "Public Opinion Polls and Perceptions of US Human Spaceflight," *Space Policy* 19 (2003): 163–75.

53. Fine examples are William E. Burrows, *This New Ocean: The Story of the First Space Age* (New York: Random House, 1998); Charles A. Murray and Catherine Bly Cox, *Apollo: The Race to the Moon* (New York: Simon & Schuster, 1989); McDougall, *Heavens and the Earth*; Logsdon, *The Decision to Go to the Moon* and *John F. Kennedy and the Race to the Moon.*

54. Alex Roland, "Barnstorming in Space: The Rise and Fall of the Romantic Era of Spaceflight, 1957–1986," in Radford Byerly, Jr., ed., *Space Policy Reconsidered* (Boulder, CO: Westview Press, 1989), 33–52.

55. Patricia Nelson Limerick, "The Adventures of the Frontier in the Twentieth Century," in James R. Grossman, ed., *The Frontier in American Culture* (Berkeley: University of California Press, 1994); Stephen J. Pyne, "Space: A Third Great Age of Discovery," *Space Policy* 4 (1988): 187–99; William H. Truettner, ed., *The West as America: Reinterpreting Images of the Frontier, 1820–1920* (Washington, DC: Smithsonian Institution Press, 1991).

56. Bogue, "Frederick Jackson Turner Reconsidered"; John W. Caughey, "The Insignificance of the Frontier in American History," *Western Historical Quarterly* 5 (1974): 4–16; Ray Allen Billington, *The American Frontier Thesis: Attack and Defense* (Washington, DC: American Historical Association, 1971); Slotkin, *The Fatal Environment,* 16, 18; William Cronon, George Miles, and Jay Gitlin, "Becoming West: Toward a New Meaning for Western History," in Cronon, Miles, and Gitlin, *Under an Open Sky: Rethinking America's Western Past* (New York: W. W. Norton, 1992), 5.

57. Discussions of the mythic frontier in the twentieth century include Chapter 21, "The Imagined West" in Richard White, *"It's Your Misfortune and None of My Own": A New History of the American West* (Norman: University of Oklahoma Press, 1991), 613–32; Robert G. Athearn, *The Mythic West in Twentieth-Century America* (Lawrence: Univer-

sity Press of Kansas, 1986); Ray Allen Billington, *America's Frontier Heritage* (New York: Holt, Rinehart and Winston, 1966); and Patricia Nelson Limerick, "The Adventures of the Frontier in the Twentieth Century," in Grossman, *The Frontier in American Culture,* 66–102. College-level history textbooks published in 1959 included Thomas D. Clark, *Frontier America: The Story of the Westward Movement* (New York: Charles Scribner's Sons, 1959), and Ray Allen Billington, *The Westward Movement in the United States* (New York: Van Nostrand, 1959).

58. Beverly J. Stoeltje, "Making the Frontier Myth: Folklore Process in a Modern Nation," *Western Folklore* 46 (1987): 235–53; Ray A. Williamson, "Outer Space as Frontier: Lessons for Today," *Western Folklore* 46 (1987): 255–67.
59. Examples from the *New York Times* included: "Still the Space Race," January 30, 1965, p. 26; "The Space Race," March 19, 1965, p. 34; "U.S. Overtakes U.S.S.R. in Space, U.S. Space Officials Say," October 2, 1965, pp. SUA4–3; "U.S. Is Ahead of Soviet in Man-in-Space Race," December 28, 1968, p. 12.
60. Dennis Jenkins, *Space Shuttle* and *X-15: Extending the Frontiers of Flight* (Washington, DC: National Aeronautics and Space Administration, 2007); Melvyn Smith, *An Illustrated History of Space Shuttle: US Winged Spacecraft X-15 to Orbiter* (Somerset, Eng.: Haynes, 1985); Heppenheimer, *Development of the Space Shuttle.*
61. Jenkins, *Space Shuttle;* Smith, *Illustrated History.*

CHAPTER 2. SPACE SHUTTLE

1. The STS-5 mission in November 1982 was the first operational mission, first commercial satellite delivery, and first four-person crew. They deployed two communications satellites.
2. The von Braun–Ley blueprint for a coherent program of human space exploration, introduced in the previous chapter, included a space station, various types of shuttlecraft, a lunar base, and a mission to Mars. Pushed to the margins by the space-race goal of going straight to the moon, this paradigm from the 1950s influenced efforts in the 1960s to define a post-Apollo space program, especially the idea of reusable spacecraft shuttling people between earth and an orbital workplace. This lineage is charted in Dwayne A. Day, "Paradigm Lost," *Space Policy* 11/3 (1995): 153–59.
3. The Space Task Group Report, *The Post-Apollo Space Program: Directions for the Future* (Washington, DC: Executive Office of the President, September 15, 1969), accessible at www.hq.nasa.gov/office/pao/History/taskgrp.html, states goals for human spaceflight, space access and applications, and exploration. See also contributions from other organizations: Executive Office of the President, *The Next Decade in Space: A Report of the Space Science and Technology Panel of the President's Science Advisory Committee* (Washington, DC: March 1970); NASA, "Goals and Objectives for America's Next Decades in Space," September 1969, and "America's Next Decades in Space: A Report for the Space Task Group," September 1969; *Space Shuttle Task Group Report* (Washington, DC: National Aeronautics and Space Administration, 1969); Task Force on Space (Charles Townes transition team report for Nixon). All of these documents are available in the NASA Historical Reference Collection.

4. T. A. Heppenheimer, *Development of the Space Shuttle, 1972–1981* (Washington, DC: Smithsonian Institution Press, 2002); Dennis R. Jenkins, *Space Shuttle: Development of the National Space Transportation System: The First 100 Missions* (North Branch, MN: Specialty Press, 2001).
5. Jenkins, *Space Shuttle,* traces this study process well. For discussions of policy influences and decision-making, see Roger D. Launius, "NASA and the Decision to Build the Space Shuttle, 1969–1972," *Historian* 57 (1994): 17–34, and "A Waning of Technocratic Faith: NASA and the Politics of the Space Shuttle Decision," *Journal of the British Interplanetary Society* 49 (1996): 49–58; also John M. Logsdon, "The Space Shuttle Decision: Technology and Political Choice," *Journal of Contemporary Business* 7/3 (1978): 13–29, and "The Decision to Develop the Space Shuttle," *Space Policy* 2 (1986): 103–19.
6. Stephen J. Garber, "Why Does the Space Shuttle Have Wings? A Look at the Social Construction of Technology in Air and Space," available at history.nasa.gov/sts1/pages/scot.html; Robert C. Truax, "Shuttles—What Price Elegance?" *Astronautics and Aeronautics,* June 1970, pp. 22–23.
7. Statement by President Nixon, January 5, 1972, NASA Historical Reference Collection, available at history.nasa.gov/stsnixon.htm, and *Public Papers of the Presidents, Richard M. Nixon, 1972* (Washington, DC: Office of the Federal Register, National Archives and Records Administration, 1974). John M. Logsdon generously shared copies of the NASA draft and near-final presidential statement found during his research in the papers of former NASA deputy administrator George M. Low, archived at Rensselaer Polytechnic Institute in Troy, New York. A rhetorical analysis of the statement appears in Linda T. Krug, *Presidential Perspectives on Space Exploration: Guiding Metaphors from Eisenhower to Bush* (New York: Praeger, 1992), 54–56.
8. Deputy Administrator George M. Low, memorandum to Assistant Administrator for Public Affairs, Project Names, November 8, 1974, NASA Historical Reference Collection, File NASA Names—Space Shuttle.
9. The NASA file "Names—Space Shuttle" in the NASA Historical Reference Collection contains many memos from 1969 to 1978 on renaming the generic space shuttle spacecraft (not the individual orbiters). The administrator forwarded several possible names to an assistant to the president just days before the decision to proceed with the space shuttle was announced. The defense of "shuttle" as a good name is found in Public Affairs Officer for Space Flight William J. O'Donnell, memorandum to Associate Administrator for External Affairs, The Space Shuttle Name Change, December 30, 1975, LeRoy E. Day Papers, Box 8, NASA Historical Reference Collection.
10. "And Now, a Real Space Shuttle Named 'Enterprise,'" *Washington Post,* September 9, 1976, p. E9.
11. "Shuttle Roll-Out Set for Sept. 17," NASA News Release No. 76–149, and "Space Shuttle Roll-Out," companion press kit, both undated, File 007952, NASA Historical Reference Collection.
12. "Space Shuttle Roll-Out," pp. 1, 3, 11.
13. "Space Shuttle Makes Its Debut," *Congressional Record* (Senate), September 21, 1976, S16327, File 007952, in NASA Historical Reference Collection. Remarks as reported in "Space Shuttle Unveiled with Star Trek Trimmings," *Washington Star,* September 18, 1976,

p. A-1, File 007952, NASA Historical Reference Collection, and in Robert Hotz, "A New Space Era," *Aviation Week & Space Technology,* October 4, 1976, p. 9.

14. The early manager of the space shuttle program, Robert F. Thompson, put the shuttle era into this historical context in a 1974 paper, "The Space Shuttle—A Future Space Transportation System." He commented at length on mobility and progress, and the role of advances in transportation in relation to economic and social improvement, citing ships, canals, railroads, aviation, and missiles as agents of change. Johnson Space Center Archives, STS General Files, Box 1. At least one editorial cartoonist captured this idea by drawing a covered wagon, locomotive, and shuttle together; Frank Miller cartoon for The Register and Tribune Syndicate, 1981, published in the *Philadelphia Inquirer* (April 26, 1981) just after the first shuttle mission.
15. *Space Shuttle,* NASA Educational Publication EP-96, June 1972; *Space Shuttle,* NASA Special Publication SP-407, 1976; *The Shuttle Era,* Space Shuttle Fact Sheet NASA-S-76–815A, NASA, March 1977. The Johnson Space Center, the engineering home of the space shuttle program, issued frequent updates of this publication. *Space Shuttle Transportation System: A Promising New Era for Earth,* Rockwell International Space Division, September 1976, and *Space Shuttle: A Promising New Era for Earth,* Rockwell International Space Division, January 1977. All of these publications are in the Shuttle Program File or Shuttle General Publications File, NASA Historical Reference Collection, or the STS General Files, Johnson Space Center Archives.
16. John Noble Wilford, "Another Small Step for Man: Shuttling into Space," *New York Times Magazine,* August 7, 1977, pp. 7ff., and "Commuting Age Dawns in Space," *New York Times,* December 30, 1979, p. DX9.
17. Roger Rosenblatt, "Space Shuttle Columbia Aiming High in '81," and Frank Tripett, "Milk Run to the Heavens," both in *TIME,* January 12, 1981; Frederick Golden, "Touchdown, Columbia," *TIME,* April 27, 1981.
18. See extensive coverage in *TIME* and *Newsweek* magazines and all major news outlets in April 1981.
19. "U.S. Car and Truck Retail Sales from 1980 to 2013," available at www.statista.com/statistics/199981/us-car-and-truck-sales-since-1951.
20. "The Space Truck," *Washington Post,* April 10, 1981, p. A18.
21. Articles in *New York Times* by Wilford and others, April 1981; headline from April 15, 1981, p. A21.
22. John Noble Wilford, "Space and the American Vision," *New York Times Magazine,* April 5, 1981, pp. 14ff. and 118ff.
23. Typically the *New York Times* ran a news article each day of each mission, several in the days just before launch and after landing, at least one article for every delay or significant problem, and occasional analytical pieces. The mission-related coverage during the 1981–85 period totaled hundreds of articles.
24. The five missions were, in 1984, the tenth (STS 41-B), featuring first flights in the Manned Maneuvering Unit; the eleventh (STS 41-C), the Solar Max observatory repair mission; the fourteenth (STS 51-A), the first satellite retrieval to return the Westar and Palapa communications satellites; and in 1985, the sixteenth (STS 51-D), another satellite delivery mission, and the twentieth (STS 51-I), to deliver three satellites and retrieve/repair another.

See *New York Times* articles by Wilford and others in January–April and November 1984, and April and August–September 1985.

25. Fifth mission (STS-5, 1982), locked wheel and flat tire; ninth mission (STS-9, 1983), aft compartment fire upon landing; sixteenth mission (STS 51-D, 1985), damaged wing and landing gear blowout.
26. "Is the Shuttle Worth Rooting For?," *New York Times,* April 9, 1981, p. A22; "Down to Earth," *New York Times,* April 14, 1981, p. A30; "What Does the 'S' in NASA Mean?," *New York Times,* November 4, 1981, p. A30.
27. "Too Fine a Machine," *New York Times,* March 31, 1982, p. A30.
28. John Noble Wilford, "Big Business in Space," *New York Times Magazine,* September 18, 1983, pp. 46–47ff.
29. Doug Marlette, *Shred This Book: The Scandalous Cartoons of Doug Marlette* (Atlanta: Peachtree Publishers, 1988), 6. See this motive at work in *The Emperor Has No Clothes: Editorial Cartoons by Marlette* (Washington, DC: Graphic Press, 1976).
30. The NASA Historical Reference Collection at NASA Headquarters in Washington, D.C., contains many cartoon files catalogued by year and topic in the series "Cartoons."
31. Pat Oliphant with Harry Katz, *Oliphant's Anthem: Pat Oliphant at the Library of Congress* (Kansas City, MO: Andrews, McMeel & Parker, 1998), 93.
32. Various scholars have examined editorial cartoons as effective keys to frames of meaning: William A. Gamson and David Stuart, "Media Discourse as a Symbolic Contest: The Bomb in Political Cartoons," *Sociological Forum* 7/1 (1992): 55–86; Edward T. Linenthal, *Symbolic Defense: The Cultural Significance of the Strategic Defense Initiative* (Urbana: University of Illinois Press, 1989); Thomas H. Bivins, "The Body Politic: The Changing Shape of Uncle Sam," *Journalism Quarterly* 63 (1987): 13–20; Roger A. Fischer, "Oddity, Icon, Challenge: The Statue of Liberty in American Cartoon Art, 1879–1986," *Journal of American Culture* 9/4 (1986): 63–81.
33. *America's Space Truck: The Space Shuttle,* in the exhibit files of the Smithsonian National Air and Space Museum. The exhibit was displayed under that title until a 1993 update to become simply *Space Shuttle.* The museum has had a space shuttle exhibit in the building on the National Mall continuously since 1981. In 2003, the shuttle test vehicle *Enterprise* went on display at the museum's new Stephen F. Udvar-Hazy Center near Washington's Dulles International Airport in Chantilly, Virginia, and in 2012 the museum relinquished *Enterprise* and permanently installed the orbiter *Discovery* at the same site.
34. Two early articles on the economics and purpose of the shuttle summarized opposing views: Lawrence Lessing, "Why the Space Shuttle Makes Sense," *Fortune,* January 1972, pp. 93–97, and Brian O'Leary, "The Space Shuttle: NASA's White Elephant in the Sky," *Bulletin of the Atomic Scientists,* February 1972, pp. 36–43. Wayne Biddle, "The Endless Countdown," *New York Times Magazine,* June 22, 1980, p. SM8, raised similar questions as the first shuttle launch approached. Since its origins, the economics of the shuttle were contested in the Office of Management and Budget and in Congress, notably by Senator Walter Mondale.
35. John Noble Wilford, "Gap Between Early Hope and Present Accomplishment Grows Large; Space Shuttle Re-evaluated," *New York Times,* May 14, 1985, p. C1.
36. Mike Toner, "It's Pay Off or Perish for the Shuttle," *Science Digest,* May 1985, pp. 64–67;

Dennis Overbye, "Success amid the Snafus," *Discover,* November 1985, pp. 53–58.; Alex Roland, "The Shuttle: Triumph or Turkey?" *Discover,* November 1985, pp. 29–49 (quotes on p. 45). Roland's critique of the human spaceflight rationale appeared as "Barnstorming in Space: The Rise and Fall of the Romantic Era of Spaceflight, 1957–1986," in Radford Byerly, Jr., ed., *Space Policy Reconsidered* (Boulder, CO: Westview Press, 1989), 33–52, and "NASA's Manned-Space Nonsense," *New York Times,* October 4, 1987, p. E23.

37. *Report of the Presidential Commission on the Space Shuttle Challenger Accident* (Washington, DC, June 6, 1986); John M. Logsdon, "The Space Shuttle Program: A Policy Failure?" *Science* 232 (1986): 1099–1105; J. A. Van Allen, untitled op-ed in *New York Times,* April 1, 1986, p. A31; Molly K. Macauley, "Rethinking Space Policy: The Need to Unearth the Economics of Space," in Byerly, *Space Policy Reconsidered,* 131–43. See also Howard E. McCurdy, "The Cost of Spaceflight," *Space Policy* 10/4 (1994): 277–89.
38. NASA Administrator James C. Fletcher, letter to Commission on Fine Arts Chairman J. Carter Brown, October 3, 1975, File 004159, NASA Historical Reference Collection; "Setting the Standard: The NEA Initiates the Federal Design Improvement Program, January 1, 1972, available at arts.gov/article/setting-standard-nea-initiates-federal-design-improvement-program; Steve (Stephen J.) Garber, "NASA 'Meatball' Logo," available at history.nasa.gov/printFriendly/meatball.htm. Executive Order 3524 issued by President Warren Harding in 1921 required the review of insignia, medals, and other designs by the Commission on Fine Arts.
39. The design firm was Danne & Blackburn Inc. of New York. NASA Administrator James C. Fletcher, letter to Commission on Fine Arts Chairman J. Carter Brown, October 3, 1975, File 004159, NASA Historical Reference Collection; "NASA Gets New Graphics Identity," NASA News Release 75–91, March 28, 1975; NASA Administrator James C. Fletcher, letter to Senator Frank E. Moss, Chairman, Committee on Aeronautical and Space Sciences, April 15, 1976; "NASA Graphic Profile Change," March 1975; "NASA Graphics Improvement Program Briefing," April 8, 1975; "The Manager's Guide to NASA Graphics Standards," NHB 1430.2; and program booklet for the Presidential Design Awards, 1984, all in File 004542, NASA Historical Reference Collection.
40. The reason for *Constitution* was not, as often mistakenly thought, to commemorate the bicentennial of the United States (1776) or the bicentennial of the Constitution (1787). Rather, it was because the first orbiter was scheduled to make its first public appearance on the anniversary date when the Constitution was signed, September 17. This intent is confirmed in several internal NASA memos, with the initial suggestion of linking the first orbiter's name to this date in a memo dated January 14, 1976. William J. O'Donnell, memo, LeRoy Day Papers, Box 8, NASA Historical Reference Collection. A memo from Public Affairs Officer for Space Flight David W. Garrett to Director, Office of Public Affairs (Name for Orbiter 102, January 26, 1978, Shuttle Chrons, Box 015–63, Johnson Space Center History Archives), suggests acting quickly "before the Trekkies learn that Enterprise will probably never go into space and they insist on something like Enterprise II." A comparable memo from Garrett on February 9, 1978, about NASA controlling the name is filed in the LeRoy Day Papers, Box 8.
41. Associate Administrator for External Affairs Herbert J. Rowe, memo and lists, to Administrator Dr. Fletcher, August 25, 1976. Associate Administrator for Space Transportation

Systems John F. Yardley, memo and lists to Director, Public Affairs, May 26, 1978, both in LeRoy Day Papers, Box 8, and File NASA Names—Space Shuttle, NASA Historical Reference Collection.

42. Arnold W. Frutkin, memo to Administrator, Dr. Frosch, June 15, 1978, File NASA Names—Space Shuttle, NASA Historical Reference Collection. Deputy Director Space Shuttle Program LeRoy E. Day, memo to Associate Administrator for Space Transportation Systems John F. Yardley, Suggested Orbiter Names, July 21, 1978; Associate Administrator for Space Transportation Systems John F. Yardley, memo to Director, Public Affairs, "Recommended Names for the Orbiters," October 27, 1978; Director of Public Affairs Robert A. Newman, memo to Deputy Administrator, November 2, 1978; all in LeRoy Day Papers, Box 8, NASA Historical Reference Collection.

43. Arnold W. Frutkin, memo to Distribution, Orbiter Names, November 3, 1978; Chief Scientist John E. Naugle, memo to Associate Administrator for External Relations, Possible Names for Shuttles, November 17, 1978; Arnold W. Frutkin, memo to Distribution, Orbiter Names, November 20, 1978; Arnold W. Frutkin, memo to Deputy Administrator Dr. Lovelace, Orbiter Names, December 11, 1978; all in File NASA Names—Space Shuttle, NASA Historical Reference Collection. "Shuttle Orbiters Named After Sea Vessels," NASA News Release No. 79–10, January 25, 1979, File 8005, NASA Historical Reference Collection; Robert Allnut and Terence Finn, memos on Recollections on Naming the Orbiters, January 3, 1983, February 14, 1983, February 21, 1983, and February 28, 1983, File 000659, NASA Historical Reference Collection. President Carter's science adviser at the time was Frank Press. A student contest yielded the name of the fifth orbiter, *Endeavour,* built to replace *Challenger.*

44. Jim Laux, memo to Terry Finn, Recommended Names for the Orbiter, November 7, 1978, File NASA Names—Space Shuttle, NASA Historical Reference Collection.

45. The NASA Space Transportation System Program Badge (Space Shuttle) was established in February 1977. NASA Management Instruction on NASA Seal, Insignia, Logotype, Program and Astronaut Badges, and Flags, File 004542, NASA Historical Reference Collection. Deputy Assistant Administrator for Public Affairs Alex P. Nagy, memo to distribution at NASA Centers, Program Insignia for Space Transportation System, February 25, 1977, Shuttle Chrons, Box 015–12, Johnson Space Center Historical Reference Collection.

46. Director (Flight Operations) Christopher C. Kraft, Jr., letter to Associate Administrator for Space Flight John F. Yardley, January 7, 1977, Shuttle Chrons, Box 015–64, Johnson Space Center Historical Reference Collection.

47. Margaret A. Weitekamp, "Softening the Orbiter: The Space Shuttle as Plaything and Icon," in Anne Collins Goodyear and Margaret A. Weitekamp, eds., *Analyzing Art and Aesthetics,* vol. 9: *Artefacts: Studies in the History of Science and Technology* (Washington, DC: Smithsonian Institution Scholarly Press, 2013), 88–103.

48. Kerry Mark Joels and Gregory P. Kennedy, *The Space Shuttle Operator's Manual* (1982; New York: Ballantine Books, repr. 1988).

CHAPTER 3. ASTRONAUTS

1. NASA's Johnson Space Center announced "NASA to Recruit Space Shuttle Astronauts" in press release 76–44 on July 8, 1976. NASA News Release No. 78–7, "NASA Selects 35 Astronaut Candidates," January 16, 1978, and "Space Shuttle Astronauts Press Conference," January 16, 1978, both in NASA Historical Reference Collection, available at mira.hq.nasa.gov/history. See also NASA Johnson Space Center Release No. 79–03, same title and date, www.nasa.gov/centers/johnson/pdf/83130main_1978.pdf. The Group 8 astronaut candidates' press conference occurred at NASA's Johnson Space Center on February 1, 1978.
2. The phrase "Reinventing the 'right stuff'" appeared in Jerry Adler, "Sally Ride: Ready for Liftoff," *Newsweek,* June 13, 1983, p. 36, an allusion to Tom Wolfe's 1979 book (and the 1983 film) *The Right Stuff,* about the culture of test pilots and Mercury astronauts.
3. Howard E. McCurdy, *Space and the American Imagination* (Washington, DC: Smithsonian Institution Press, 1997); Michael J. Neufeld, ed., *Spacefarers: Images of Astronauts and Cosmonauts in the Heroic Era of Spaceflight* (Washington, DC: Smithsonian Institution Scholarly Press, 2013).
4. Joseph Campbell, *The Hero with a Thousand Faces,* 2nd ed. (Princeton, NJ: Princeton University Press, 1968); Carl G. Jung, *Man and His Symbols* (London: Aldus Books, 1964); Leonie Cooper, "Imagining Astronauts," in Emma McRae and Sarah Tutton, eds., *Star Voyager: Exploring Space on Screen* (Melbourne: Australian Centre for the Moving Image, 2011), 45–53; Roger D. Launius, "Heroes in a Vacuum: The Apollo Astronaut as Cultural Icon," *Florida Historical Quarterly* 87/2 (2008): 174–209.
5. Joseph D. Atkinson, Jr., and Jay M. Shafritz, *The Real Stuff: A History of NASA's Astronaut Recruitment Program* (New York: Praeger, 1985), 18–53 (Chapter 3, "Selection of Pilots: The Original Seven (Group I)"); Loyd S. Swenson, Jr., James M. Grimwood, and Charles C. Alexander, *This New Ocean: A History of Project Mercury* (Washington, DC: National Aeronautics and Space Administration, 1966; repr. 1999), 160.
6. Matthew H. Hersch, *Inventing the American Astronaut* (New York: Palgrave Macmillan, 2012); Matthew H. Hersch, "'Capsules Are Swallowed': The Mythology of the Pilot in American Spaceflight," in Neufeld, *Spacefarers,* 35–55; Dario Llinares, *The Astronaut: Cultural Mythology and Idealised Masculinity* (Cambridge: Cambridge Scholars Publishing, 2011).
7. African American air force test pilot Edward J. Dwight, Jr., participated in pre-astronaut training in 1963 but was not selected as a candidate, and another African American, Robert H. Lawrence, died in a 1967 flight training accident while preparing for the U.S. Air Force astronaut program. For early interest in women astronauts and the Lovelace Clinic female pilots research program, see Margaret Weitekamp, *Right Stuff, Wrong Sex: America's First Women in Space Program* (Baltimore: Johns Hopkins University Press, 2004).
8. Tom Wolfe, *The Right Stuff* (New York: Farrar, Straus and Giroux, 1979).
9. Hersch, in *Inventing the American Astronaut* and "Capsules Are Swallowed," examines in detail the uneasy social and professional relations between pilot and scientist astronauts during the Apollo era.
10. *TIME,* December 6, 1968; January 3, 1969; and July 25, 1969.
11. Atkinson and Shafritz, Chapters 5 and 6 in *The Real Stuff,* 87–179. Kim McQuaid, "Race, Gender, and Space Exploration: A Chapter in the Social History of the Space Age,"

Journal of American Studies, 41/2 (2007): 401–34, details the fraught effort to establish equal opportunity within NASA.

12. "The Statement by President Nixon, January 5, 1972," available at history.nasa.gov/stsnixon.htm. NASA public relations materials such as *Space Shuttle* (a 1976 booklet) and *The Shuttle Era* (a 1977 fact sheet) linked the crew to the versatility of the spacecraft, and even mentioned the possibility of passengers.
13. David J. Shayler and Colin Burgess, *NASA's Scientist-Astronauts* (Chichester, UK: Springer-Praxis, 2007).
14. Amy E. Foster, *Integrating Women into the Astronaut Corps: Politics and Logistics at NASA, 1972–2004* (Baltimore: Johns Hopkins University Press, 2011). On the shuttle announcement and women, see 77–79.
15. Atkinson and Shafritz, Chapter 6, "Selection of Pilots and Mission Specialists (Group VIII)," in *The Real Stuff*, 133–79.
16. "NASA Gathers Data for Setting Female Astronaut Criteria," NASA Johnson Space Center Release No. 76–65, October 19, 1976; Harold Sandler and David L. Winter, *Physiological Responses of Women to Simulated Weightlessness: A Review of the Significant Findings of the First Female Bed-Rest Study*, NASA SP-430 (Washington, DC: NASA, 1978); Foster, *Integrating Women into the Astronaut Corps*, 79–81, 85.
17. "NASA to Recruit Space Shuttle Astronauts," NASA Release No. 76–44, July 8, 1976; "Astronauts Wanted: Women, Minorities Are Urged to Apply," *New York Times*, July 8, 1976, p. A23. The next recruiting announcements more directly targeted minority groups: "NASA to Recruit Space Shuttle Astronauts," NASA Release No. 79–050, August 1, 1979; "Minorities Are Encouraged to Apply for Astronaut Program," NASA Johnson Space Center Release No. 79–057, September 12, 1979; "Hispanics Are Encouraged to Apply for Astronaut Program," NASA Release No. 79–056, September 12, 1979.
18. Atkinson and Shafritz, *The Real Stuff*, 154–57.
19. Sara Sanborn, "Sally Ride, Astronaut: The Whole World Is Watching," *Ms.*, June 1983, p. 48; Michael Cassutt, "Mr. Inside [George Abbey]," *Air & Space*, August 2011.
20. "More than 8,000 Apply for Space Shuttle Astronaut Program," NASA Release No. 77–145, July 15, 1977; "Space Shuttle Astronauts Press Conference," January 16, 1978, NASA Historical Reference Collection, available at mira.hq.nasa.gov/history. "First Women, Blacks on Astronaut List," *Washington Post*, January 16, 1978, p. A2; "35 Chosen Astronaut Candidates: Six Are Women and Three Blacks," *New York Times*, January 17, 1978, p. 14; "NASA Announces Selection of 35 Shuttle Astronauts," *Washington Post*, January 17, 1978, p. A3.
21. "Space Shuttle Astronauts Press Conference," January 16, 1978, NASA Historical Reference Collection, available at mira.hq.nasa.gov/history.
22. Earl Lane of *Newsday*, published as "Skylab [*sic*] Promises Humanized Astronauts: Class of 35 Now in Training Will Be First True Career Spacemen," in *Los Angeles Times*, November 12, 1978, p. F1.
23. Jennifer Ross-Nazzal, "You've Come a Long Way, Maybe: The First Six Women Astronauts and the Media," in Neufeld, *Spacefarers*, 175–201; Astronaut trainees photo, *New York Times*, February 1, 1978, p. A1; "NASA Picks Six Women Astronauts with the Message: You're Going a Long Way, Baby," *People*, February 6, 1978, pp. 28–30; Mary Lu Abbott, "Space Women: Men Astronauts Take Backseat in NASA's Newest Crew Lineup,"

Houston Chronicle, February 1, 1978, pp. 12–13. Journalist Thomas O'Toole favored the term "new breed" for the shuttle astronauts in his articles "Thirty-Five New Guys; They Don't Hire Astronauts Like They Used To," *Washington Post Sunday Magazine,* July 20, 1980, pp. 14ff., and "An Old Fever Burns in New Generation of Astronauts," *Los Angeles Times,* September 3, 1980, pp. K1–3; "U.S. Spacewomen Are Ready," *Chicago Tribune,* March 23, 1980, p. 4; others, such as Paul Recer, "Enter the Specialist—a New Breed of Astronaut," *U.S. News & World Report,* February 23, 1981, pp. 60, 62, also used the term.

24. Editorial cartoons and comics: "The Woman's Touch," *Los Angeles Herald Examiner,* January 19, 1978; Jim Gottenberg, ". . . lipstick!," *Today,* January 23, 1978, p. 6A; Lichty & Wagner, ". . . I'm not going to do the dishes!" *Washington Post,* February 23, 1978, p. D7; Oliphant, ". . . your duties as a woman astronaut," *Washington Star,* February 1, 1978, p. A-19; Stayskal, ". . . my wife insisted on going along, too!," *Chicago Tribune,* January 25, 1978; Lichty & Wagner, "My wife made me promise . . . ," *Washington Post,* March 13, 1978, p. C20; Richard Locher, "Anybody got any laundry?," *Chicago Tribune,* June 24, 1983, p. 10.
25. "U.S. Spacewomen Are Ready." See Henry S. F. Cooper, Jr., *Before Lift-Off: The Making of a Space Shuttle Crew* (Baltimore: Johns Hopkins University Press, 1987), for a vivid account of crew training, and Foster, *Integrating Women into the Astronaut Corps,* for the first women's experiences in training.
26. "Women Picked in Astronaut Program," *Washington Post,* September 1, 1979, p. B4; O'Toole, "An Old Fever Burns in New Generation of Astronauts."
27. William K. Stevens, "New Generation of Astronauts Poised for Shuttle Era," *New York Times,* April 6, 1981, p. A1; Recer, "Enter the Specialist"; Paul VanSlambrouck, "Space Shuttle Allows NASA to Tap New Breed of Astronaut-Specialist," *Christian Science Monitor,* April 8, 1981, p. 3.
28. Ross-Nazzal, "You've Come a Long Way, Maybe."
29. O'Toole, "An Old Fever Burns in New Generation of Astronauts."
30. Biographical profiles of payload specialists Charles D. Walker, Sultan Salman Abdulazziz al Saud, Rodolfo Neri Vela, Senator Jake Garn, and Representative Bill Nelson are available in the category "Former Astronauts or International Astronauts" at www.jsc.nasa.gov/Bios.
31. Hersch, *Inventing the American Astronaut,* 149–51.
32. "The Citizen Astronauts," *New York Times,* September 7, 1983, p. A22; "dear nasa," *New York Times,* December 19, 1983, p. A18; "Process Begins to Select First Journalist in Space," Associated Press, October 23, 1985; "Journalists Selected as 100 Semifinalists for a Space Mission," *New York Times,* April 17, 1986, p. B11. NASA and CNN later made an agreement to send veteran science and technology correspondent Miles O'Brien on a shuttle mission, but the *Columbia* tragedy put an end to such an opportunity. See O'Brien's profile at www.nasa.gov/offices/nac/members/Obrien-bio_prt.htm.
33. Astronaut Clayton C. Anderson's memoir, *The Ordinary Spaceman* (Lincoln: University of Nebraska Press, 2015).
34. Robert K. Merton, *The Sociology of Science* (Chicago: University of Chicago Press, 1973): Michael T. Kaufman, "Robert K. Merton, Versatile Sociologist and Father of the Focus Group, Dies at 92," *New York Times,* February 24, 2003; Mary Rourke, "Robert K.

Merton, 92; Pioneering Sociologist Coined 'Role Model' and Other Popular Terms," *Los Angeles Times,* March 2, 2003; Gerald Holton, "Biographical Memoir: Robert K. Merton," *Proceedings of the American Philosophical Society,* December 2004, 506–17.

35. "Three Shuttle Crews Announced," NASA News Release No. 82–023, April 19, 1982; "Woman and Black Named for Shuttle Missions," *New York Times,* April 20, 1982, p. C3; "Woman, Black Set for 'Firsts' on the Shuttle," *Washington Post,* April 30, 1982, p. A24.
36. Foster, *Integrating Women into the Astronaut Corps;* Lynn Sherr, *Sally Ride, America's First Woman in Space* (New York: Simon & Schuster, 2014); Ross-Nazzal, "You've Come a Long Way, Maybe"; Thomas O'Toole, "Sally Ride Soars at Her First News Session," *Washington Post,* May 25, 1983, p. A5; William K. Stevens, "Feminism Paved Astronaut's Way," *New York Times,* May 2, 1982, p. 70; Robert Reinhold, "Americans in Space: Women Are Ready," *New York Times,* June 7, 1983, pp. C1 and C9; Sanborn, "Sally Ride, Astronaut: The World Is Watching," cover and pp. 45–48; "Space Woman" cover and Jerry Adler, "Sally Ride: Ready for Liftoff," *Newsweek,* June 13, 1983, pp. 37–43; "O What a Ride!" cover and Michael Ryan, "A Ride in Space," *People Weekly,* June 20, 1983, p. 82ff.
37. Cover and "Careers Behind the Launchpad," *Black Enterprise,* February 1983, pp. 59–60, 64; Clarence Waldron, "Guy Bluford: Black Astronaut Makes First Space Mission," *Jet,* September 5, 1983, cover and pp. 21–24; Penelope McMillan, "Black Shuttle Astronaut Sees Self as 'Role Model,'" *Los Angeles Times,* July 14, 1983, p. 1; Bill Prochnau, "Guy Bluford: NASA's Reluctant Hero," *Washington Post,* August 21, 1983, p. A1; Howard Benedict, "Shuttle Breaks the Racial Barrier," *Chicago Tribune,* August 29, 1983, p. D12; Thomas O'Toole, "Shuttle Poised for Milestone Mission," *Washington Post,* August 29, 1983, p. A1; William J. Broad, "First U.S. Black in Space," *New York Times,* August 31, 1983, p. B6; photo caption "Challenger, Bluford: Another first" in Sharon Begley, "NASA's Nighttime Spectacular," *Newsweek,* September 5, 1983, p. 69; "Black Astronaut Returns to Earth a 'Space Hero,'" *Jet,* September 26, 1983, p. 5.
38. Cover stories featuring shuttle astronauts were relatively rare. *Newsweek* featured only John Young and Robert Crippen (April 27, 1981), Sally Ride (June 13, 1983), Shannon Lucid (October 7, 1996), and John Glenn (October 26, 1998). *TIME* featured Michael Foale (November 3, 1997) and John Glenn (August 17, 1998). *People* put Sally Ride (June 20, 1983), Christa McAuliffe (February 10, 1986), the last *Columbia* crew (February 17, 2003), and Lisa Nowak (February 19, 2007) on covers. *People* carried inside feature stories on the selection of the first women astronauts (February 6, 1978), the first astronaut marriage uniting classmates Rhea Seddon and Robert "Hoot" Gibson (September 7, 1981), the first female shuttle pilot and commander Eileen Collins (May 11, 1998); chose Mae Jemison as one of "The 50 Most Beautiful People in the World" (May 3, 1993); and named Shannon Lucid one of "The 25 Most Intriguing People '96" (December 30, 1996) and John Glenn among "The 25 Most Intriguing People '98" (December 28, 1998).
39. Example articles in *Jet* include "The New Astronauts," March 9, 1978, cover and p. 23; "Space Shuttle Gets First Black Pilot," May 20, 1985, p. 22; "L.A. Doctor Named First Black Female Astronaut," June 22, 1987, p. 5; "Dr. Mae Jemison Becomes First Black Woman in Space," September 14, 1992, cover and p. 34; "Houston Physician Is First Black to Walk in Space," February 1995, p. 47; "Two Blacks on Discovery Launch," December 25, 2006–January 1, 2007, p. 64. Example articles in *Ebony* include "Space Trio: New Faces Among Shuttle Crew," March 1979, pp. 54–62; "The Future-Makers," August

1985, pp. 62–63; "First Black Space Commander," May 1990, pp. 78–82. Feature articles on Mae Jemison include Warren E. Leary, "A Determined Breaker of Boundaries: Mae Carol Jemison," *New York Times,* September 13, 1992, p. 42; Marilyn Marshall, "Child of the '60s Set to Become First Black Woman in Space," *Ebony,* August 1989, pp. 50–55, and "Close-Up: A New Star in the Galaxy," *Ebony,* December 1992, p. 122; "Mae Jemison," *People,* May 3, 1993, p. 145.

40. Example *New York Times* editorials include "Stranded on the Sea of Tranquillity [*sic*]: Lumbering Humans Are the Real Limit on NASA's Future," July 20, 1984, p. A26; "The Wrong Stuff," November 21, 1984, p. A20; "Star Truck," June 23, 1986, p. A14; and many others in the 1990s.

41. "President Obama Names Presidential Medal of Freedom Recipients," The White House, Office of the Press Secretary, August 8, 2013, available at www.whitehouse.gov; "Remarks by the President at Presidential Medal of Freedom Ceremony," The White House, Office of the Press Secretary, November 20, 2013, available at www.whitehouse.gov. The official notice of Sally Ride's death appeared on her company website, sallyridescience.com, on July 23, 2012; the last line identified Tam O'Shaughnessy as her partner of twenty-seven years. See tributes and remembrances posted on sallyridescience.com and Joe Garofoli, "In Death, Sally Ride Inspires a New Audience," *Seattle Times,* July 25, 2012.

42. "The Tragedy of Lisa Nowak," *New York Times,* February 8, 2007, p. A20; Jeffrey Kluger, "Houston, She's Got Some Problems," and Hillary Hylton, "Why Astronauts Don't Like Shrinks," both in *TIME,* February 8, 2007; John Schwartz, "For Astronauts and Their Families, Lives with Built-In Stress," *New York Times,* February 9, 2007, p. A15; Erica Goode, "Wanted in Space: Gregarious Loners Who Take Risks, Cautiously," *New York Times,* February 11, 2003, p. F1; "Astronaut Love Triangle: Out of This World," *People,* February 19, 2007, cover and pp. 58–63; "Rapid Descent," *Newsweek,* February 19, 2007.

43. Gregory's missions were STS-51B/Spacelab 3 (1985), STS-33 (1989), and STS-44 (1991); Bolden's were STS-61C (1986), STS-31 (1990), STS-45 (1992), and STS-60 (1994). Collins's missions were STS-63 (1995), STS-84 (1997), STS-93 (1999), and STS-114 (2005).

44. Kathy Sawyer, "A Number of 'Firsts' on Flight: Female Space Pilot Joins All-Male Club," *Washington Post,* February 4, 1995, p. A3; "Woman Leads Shuttle Crew for History-Making Mission," *New York Times,* July 17, 1999, p. A10; Beth Dickey, "Woman's Work: Space Commander," *New York Times,* July 24, 1999, p. A10; Jim Yardley, "A Shuttle Leader Is Ready 'to Go Fly Again,'" *New York Times,* February 7, 2003, p. A1; John Schwartz, "To Return Shuttle to Space, NASA Calls on Cool Leader," *New York Times,* April 17, 2005, p. 21; Anna Quindlen, "The Barriers, and Beyond," *Newsweek,* August 22, 2005. Susan Still Kilrain piloted two shuttle missions, and Pamela A. Melroy piloted two missions and commanded one.

45. Images like the ones discussed in this chapter abound in NASA's vast online image archives and are too numerous to cite specifically.

46. "Faster than a Speeding Space Shuttle," CollectSpace.com, June 13, 2006, available at www.collectspace.com/news/news-061306a.html.

47. The Hubble Space Telescope deployment and servicing missions were STS-31 (1990), STS-61 (1993), STS-82 (1997), STS-103 (1999), STS-109 (2002), and STS-125 (2009); Kathy Sawyer, "Mission Has Big Risks for NASA; Agency's Reputation May Ride with Crew to Repair Telescope," *Washington Post,* November 28, 1993, p. A1. All shuttle crews

except three from May 2000 (STS-101) until the last in July 2011 (STS-135) built or supplied the International Space Station.

48. Valerie Neal, "Bringing Spaceflight Down to Earth: Astronauts and *The* IMAX *Experience*®," in Neufeld, *Spacefarers,* 149–74. The seven "filmed in space by the astronauts" IMAX films were *The Dream Is Alive* (1985), *Blue Planet* (1990), *Destiny in Space* (1994), *Mission to Mir* (1997), *Space Station 3D* (2002), *Hubble 3D* (2010), and *A Beautiful Planet* (2016).
49. Vincent Canby, "'Big Screen' Takes on New Meaning," *New York Times,* April 19, 1987; Glenn Whipp, "Hubble 3D," *Los Angeles Times,* March 19, 2010; Roger Ebert, "Hubble 3D: A Journey into Time and Space," *Chicago Sun Times,* April 21, 2010; Justin Chang, "Hubble 3D," *Variety,* March 14, 2010.
50. Lane Wallace, "Are Astronauts Heroes?" *The Atlantic,* July 2009.
51. *People* magazine gave this explanation for featuring only Christa McAuliffe on the cover and in its first story about the *Challenger* tragedy: "Chosen to represent each of us on that mission, she epitomized the person for whom we feel a special affinity: the ordinary citizen doing extraordinary things." Billie Kajunski, "Publisher's Letter," *People,* March 3, 1986. See also "The Death of a School Teacher," *New York Times,* January 29, 1986, p. A22.
52. *People,* February 17, 2003.
53. *The Right Stuff* author Tom Wolfe observed, "The last role in the world NASA had in mind for Christa McAuliffe and the rest of the *Challenger* crew was that of pioneer or hero," in "Everyman vs Astropower," *Newsweek,* February 2, 1986, p. 40.
54. Marita Sturken, *Tangled Memories: The Vietnam War, the AIDS Epidemic, and the Politics of Remembering* (Berkeley: University of California Press, 1997), 33–37; Ed Magnuson, "Space: They Slipped the Surly Bonds of Earth to Touch the Face of God," *TIME,* February 10, 1986, cover story.
55. *TIME* and *Newsweek* issues dated February 10, 1986, and February 10, 2003; in contrast, *People,* February 17, 2003, featured the *Columbia* crew's walkout photo. See Robert Hariman and John Louis Lucaites, *No Caption Needed: Iconic Photographs, Public Culture, and Liberal Democracy* (Chicago: University of Chicago Press, 2007).
56. Kristin Ann Hass, *Carried to the Wall: American Memory and the Vietnam Veterans Memorial* (Berkeley: University of California Press, 1998), explicates this participatory impulse. See also Kendall R. Phillips, ed., *Framing Public Memory* (Tuscaloosa: University of Alabama Press, 2004), especially the essay by Edward S. Casey, "Public Memory in Place and Time," 17–44.
57. Robin Toner, "Space Museum Becomes a Memorial," *New York Times,* January 30, 1986, p. A18, and Mary Combs, "Shuttle Disaster Mourners Flock to NASM," *The [Smithsonian] Torch* 86 (1986): 1. Scholarly literature about these practices includes Edward T. Linenthal, *The Unfinished Bombing: Oklahoma City in American Memory* (New York: Oxford University Press, 2001); C. Allen Haney et al., "Spontaneous Memorialization: Violent Death and Emerging Mourning Rituals," *Omega: The Journal of Death and Dying* 35/2 (1997): 159–71; and Sylvia Grider, "Spontaneous Shrines: A Modern Response to Tragedy and Disaster," *New Directions in Folklore* 5 (2001), available at www.temple.edu/isllc/newfolk/shtrines/html. News reports of spontaneous observances included "Communities Across Nation Grope for Ways to Express Grief" and "A New Tragedy Calls Back the Sorrows of the Past," both in *Washington Post,* January 30, 1986, pp. A1 and B1.

58. J. Bodnar, *Remaking America: Public Memory, Commemoration, and Patriotism in the Twentieth Century* (Princeton, NJ: Princeton University Press, 1992); Linenthal, *Unfinished Bombing;* J. B. Gardner and S. M. Henry, "September 11 and the Mourning After: Reflections on Collecting and Interpreting the History of Tragedy," *Public Historian* 24 (2002): 37–52; Erica Doss, "Death, Art, and Memory in the Public Sphere: The Visual and Material Culture of Grief in Contemporary America," *Mortality* 7 (2002): 63–82.
59. Mary E. Stuckey, *Slipping the Surly Bonds: Reagan's* Challenger *Address* (College Station: Texas A&M University Press, 2006), 5–13, and *The President as Interpreter-in-Chief* (Chatham, NJ: Chatham House, 1991); R. W. Apple, "President as Healer," *New York Times,* January 29, 1986, p. A7.
60. President Ronald W. Reagan, "Explosion of the Space Shuttle *Challenger,* Address to the Nation on January 28, 1986," available at history.nasa.gov/reagan12886.html and in *Public Papers of the Presidents of the United States: Ronald Reagan, 1986,* Book 1 (Washington, DC: Office of the Federal Register, National Archives and Records Administration, 1988), 95; Peggy Noonan, *What I Saw at the Revolution* (New York: Random House, 1990), 252–59; Stuckey, *Slipping the Surly Bonds,* 80–81; Linda T. Krug, *Presidential Perspectives on Space Exploration: Guiding Metaphors from Eisenhower to Bush* (New York: Praeger, 1991), Chapters 4 and 5 on Reagan, 67–89; Richard Cohen, "Reagan Made It Better," *Washington Post,* January 31, 1986, p. A19. Ronald Reagan, "Remarks at the Memorial Service for the Crew of the Space Shuttle *Challenger,*" Houston, Texas, January 31, 1986, *Public Papers of the Presidents of the United States: Ronald Reagan, 1986,* Book 1, 109–11.
61. George W. Bush, "Address to the Nation on the Loss of Space Shuttle *Columbia,*" February 1, 2003, and "Remarks at a Memorial Service for the STS-107 Crew of Space Shuttle *Columbia* in Houston, Texas," February 4, 2003, in *Public Papers of the Presidents of the United States: George W. Bush, 2003,* Book 1 (Washington, DC: Office of the Federal Register, National Archives and Records Administration, 2005), 119–20, 126–27.
62. Some examples of these cartoons are Doug Marlette's eagle head profile with a single weeping tear (*Charlotte Observer,* 1986); Steve Greenberg's spacesuited Uncle Sam with helmet off, slumped over a table, one hand covering his eyes in grief and the other resting on a small space shuttle (*Seattle Post-Intelligencer,* January 29, 1986); Daryl Cagle's weeping child sitting in a toy shuttle (Slate.com, February 2, 2003); and Sean Leahy's seven falling stars, one a star of David, streaking across the frame like *Columbia* debris (*Courier-Mail,* Brisbane, Australia); and a variety of constellations in the shape of a shuttle. Many such cartoons are archived at cagle.slate.msn.com and can be viewed by using the search terms "Challenger" or "Columbia."
63. See, for example, in the *New York Times,* William J. Broad, "Thousands Watch a Rain of Debris," January 29, 1986, p. A1; Philip M. Boffey, "Troubling Questions," January 29, 1986, p. A7; David E. Rosenbaum, "Should U.S. Continue to Send People into Space?," January 30, 1986, p. A18; William J. Broad, "Astronauts Active in Quest for Answers About Disaster," February 18, 1986, p. C3; Robert Reinhold, "Astronauts' Chief Says NASA Risked Life for Schedule," March 9, 1986, p. 1; James Glanze, "Bureaucrats Stifled Spirit of Adventure, NASA's Critics Say," February 18, 2003, p. A1; John Noble Wilford, "The Allure of Mars Grows as U.S. Searches for New National Goal," March 24, 1987, p. C1.

64. In the *New York Times,* "The Challenge Beyond Challenger," January 31, 1986, p. A30; Russell Baker, "Nobody Seemed to Worry," February 1, 1986, p. 27; "Risk and Routine in Space," February 7, 1986, p. A34; "NASA's Future in Space," April 3, 1986, p. A26; "Star Truck," June 23, 1986, p. A14; "Drifting Back to Space," July 29, 1986, p. A22; Carl Sagan, "It's Time to Go to Mars," January 23, 1987, p. A27; and others in early 1986; "The Call of Distant Worlds," February 9, 2003, p. WK14; "The Challenge Ahead in Space," July 6, 2003, p. WK8.
65. In the *New York Times,* see these articles by John Noble Wilford: "The Shuttle Seems Set to Declare Dividends," December 29, 1985, p. E16; "Faith in Technology Is Jolted, but There Is No Going Back," January 29, 1986, p. A7; "The Challenger's Fate, the Shuttles' Future," February 2, 1986, p. E1; "After the Challenger: America's Future in Space," March 16, 1986, pp. 85ff.; "The Allure of Mars Grows as U.S. Searches for New National Goal," March 24, 1987, p. C1; "2nd Shuttle Disaster Creates Uncertain Future for Program," February 2, 2003, p. 33.
66. For example, Robert C. Cowan, "Shuttle Tragedy Stuns US, Highlights Dangers," *Christian Science Monitor,* January 29, 1986, p. 1; William J. Broad, "Astronauts Discover Policy Role Is Limited," *New York Times,* January 2, 1987, p. C1: James Reston, Jr., "The Astronauts After Challenger," *New York Times Sunday Magazine,* January 25, 1987, pp. 46ff.
67. D'Vera Cohn, "Challenger Crew Is Honored at Arlington," *Washington Post,* March 22, 1987, p. B3; "NASA Dedicates Space Shuttle Columbia Memorial," NASA News Release 04–049, February 2, 2004; "Astronaut Memorial to Be Built," NASA News Release 86–38, April 2, 1986; Jerry Adler, "Putting Names in the Sky: 'Space Mirror' Honors America's Dead Astronauts," *Newsweek,* May 13, 1991, p. 69; Carole Blair and Neil Michel, "Commemorating in the Theme Park Zone: Reading the Astronauts Memorial," in Thomas Rosteck, ed., *At the Intersection: Cultural Studies and Rhetorical Studies* (New York: Guilford Press, 1999), 29–83.
68. "7 Astronauts' Survivors Open $1 Million Foundation Drive," *New York Times,* September 24, 1986, p. A29.
69. The United States Astronaut Hall of Fame building closed in late 2015 for its contents and programs to be incorporated into a new entity, Heroes and Legends, at the Kennedy Space Center Visitor Complex.
70. Marcia Dunn, "Astronauts' Right Stuff Is Different Now," *Los Angeles Times,* October 13, 1991, p. A2.
71. James Oberg, "Unpublished NASA Report Says All-Woman Flight Isn't Necessary," UPI, March 8, 1999; "The Right Sex or the Right Stuff?" *Newsweek,* April 5, 1999; Kathy Sawyer, "Thelma and Louise in Space? Idea of All-Female Crew Could Redefine 'Unmanned' in NASA Speak," *Washington Post,* April 11, 1999, and "NASA Floats Idea of an All-Female Shuttle Mission," *Seattle Post Intelligencer,* April 13, 1999; Marcia Dunn, "NASA Considering All Female Crews," Associated Press, April 11, 1999.
72. Those born abroad as U.S. citizens or émigrés who became citizens included Fernando Caldeiro, Gregory Chamitoff, Franklin Chang-Diaz, Kalpana Chawla, Michael Foale, Gregory Johnson, Michael Lopez-Alegría, Shannon Lucid, Carlos Noriega, Piers Sellers, and Andrew Thomas. Sally Ride's comment from an interview with the *New York Times* appeared in William K. Stevens, "Feminism Paved Astronaut's Way," *New York Times,*

May 2, 1982, p. 70; in "I'm One of Those People," *New York Times,* June 18, 1983, p. 22; and elsewhere.

73. Astronaut biographical data pages are found at www.jsc.nasa.gov/Bios. These remarks are based on observation of astronaut interactions with visitors of all ages, at the Smithsonian National Air and Space Museum, where shuttle astronauts often appear in informal programs. These shuttle-era astronauts published memoirs by 2015: Joe Allen, Clayton Anderson, Chris Hadfield, Bernard Harris, José Hernández, Tom Jones, Jerry Linenger, Mike Mullane, Jerry Ross, Winston Scott, Rhea Seddon, and Don Thomas.

CHAPTER 4. SCIENCE

1. "The Flying Ph.D.s," *Newsweek,* October 7, 1996, available at www.newsweek.com/1996/10/06/the-flying-ph-d-s.print.html.
2. Hans Mark, *The Space Station: A Personal Journey* (Durham, NC: Duke University Press, 1987). See especially Chapters 11 and 12.
3. Mark, *The Space Station,* 132.
4. Mark, *The Space Station,* 140–41.
5. David Shapland and Michael Rycroft, *Spacelab: Research in Earth Orbit* (Cambridge: Cambridge University Press, 1984).
6. Spacelab 1 mission scientists Charles R. Chappell and Karl Knott emphasized the importance of direct communication and collaboration in "The Spacelab Experience: A Synopsis," *Science* 225 (1984): 163–65.
7. Douglas R. Lord, *Spacelab: An International Success Story* (Washington, DC: National Aeronautics and Space Administration, 1987), 59.
8. Deputy Associate Administrator (Wernher von Braun), memorandum to Associate Administrator for Manned Space Flight, "Request for Development of a Presentation on the Space Shuttle Sortie Mode of Operation from the Principal Investigators Point of View," February 24, 1971, Record Group 255, 73A-772, Box 2, National Archives, Washington National Records Center.
9. Lord, Chapter 2, "Birth of a Concept, 1960–1973," in *Spacelab,* 35–60; Space Science Board, National Research Council, *Scientific Uses of the Space Shuttle* (Washington, DC: National Academy of Sciences, 1974).
10. For example, C. R. Chappell, "Spacelab Science," AIAA paper 79–3062, American Institute of Aeronautics and Astronautics, 1979.
11. See Charles Walker biographical data at www.jsc.nasa.gov/Bios/PS/walker.html.
12. "Spacelab," NASA Facts, Release 77–125, June 1977; *Spacelab,* 13M883 (Huntsville, AL: National Aeronautics and Space Administration Marshall Space Flight Center, 1983); *Spacelab 1* (Huntsville, AL: NASA Marshall Space Flight Center, 1983); Walter Froehlich, *Spacelab: An International Short-Stay Orbiting Laboratory,* EP-165 (Washington, DC: National Aeronautics and Space Administration, October 1983).
13. The recruitment/selection process and the role of mission specialist astronauts is discussed in Chapter 3. Scientist astronauts recruited in 1967 and still available for shuttle missions were Joseph Allen, Anthony England, Owen Garriott, Karl Henize, William Lenoir, Story Musgrave, Robert Parker, and William Thornton. The thirty-five-member

astronaut class of 1978 included twenty scientists (mission specialists). Thomas O'Toole, "New Breed of Astronauts Takes the Leap," *Washington Post Service* in *Pittsburgh Press,* September 3, 1980, p. C-1; Paul van Slambrouk, "Space Shuttle Allows NASA to Tap New Breed of Astronaut-Specialist," *Christian Science Monitor,* April 8, 1981; Howard Benedict, "New Breed of Astronauts Is Cross-Section," Associated Press, April 12, 1981; Robert Reinhold, "Shuttle Mission Puts Focus on Research Crewmen," *New York Times,* November 13, 1982, p. 10.

14. NASA Deputy General Counsel, memorandum to [Noel Hinners] (Associate Administrator for Space Science), "Conditions of Employment for Payload Specialists for First Spacelab Payload," July 6, 1977, distributed by Robert A. Kennedy (Director, Spacelab Payloads Program), July 25, 1977, Shuttle Chrons, Box 015–22, Johnson Space Center Archives. Also, Draft NASA Management Instruction to establish policy and process for Payload Specialists, August 17, 1977, Shuttle Chrons, Box 015–24, NASA Johnson Space Center Archives; Deputy Assistant Administrator for Procurement, memorandum distributing revised draft, "Proposed NMI re Selection of Payload Specialists," September 20, 1977, Shuttle Chrons, Box 015–34, NASA Johnson Space Center Archives; AC/Special Assistant to the Director, memorandum to AC/Technical Assistant to the Director (named Hooks), "Payload Specialist Activities—a Recent History and Some Suggestions," September 6, 1977, Shuttle Chrons, Box 015–32, NASA Johnson Space Center Archives.
15. Noel Hinners (Associate Director for Space Science), telegram to Christopher Kraft (Director, Johnson Space Center), "Science Astronaut Mission Specialist Availability for Spacelab 1 and 2," August 24, 1977, Shuttle Chrons, Box 015–31, NASA Johnson Space Center Archives, and letters exchanged between Christopher Kraft (Director, Johnson Space Center) and William R. Lucas (Director, Marshall Space Flight Center), June 20 and July 12, 1977, Shuttle Chrons, Box 015–16, NASA Johnson Space Center Archives.
16. "NASA Restudies Process for Picking Shuttle Crew," *Aviation Week & Space Technology,* June 19, 1978, p. 31.
17. Department of Housing and Urban Development, *Hearings before a Subcommittee of the Committee on Appropriations, House of Representatives, Part 4: National Aeronautics and Space Administration,* 96th Congress, 1st Sess. (March 7, 1979), 156–57.
18. "Policy Paper: Expanding Opportunities to Fly Payload Specialists on the STS," appended to letters from NASA administrator James M. Beggs to key White House staff, October 7, 1982, OS Box 6, Folder 1089000, Ronald Reagan Presidential Library; "The Role of the Payload Specialist in the Space Transportation System for NASA or NASA-Related Payloads," NASA pamphlet, no date.
19. Memos exchanged between Noel Hinners (Associate Administrator for Space Science) and Christopher Kraft (Director, Johnson Space Center), September 9, September 28, and December 20, 1977, Shuttle Chrons, Box 015–55, NASA Johnson Space Center Archives.
20. Shapland and Rycroft, *Spacelab: Research,* 102.
21. Lord, *Spacelab,* 360; Shapland and Rycroft, *Spacelab: Research,* 134–35, 177–80.
22. Numerous Spacelab mission news reports by John Noble Wilford and William Broad appeared in the *New York Times* (October–December 1983, especially late November–mid-December). Others who covered Spacelab missions included Warren E. Leary in the *New York Times* and Thomas O'Toole and Kathy Sawyer in the *Washington Post.*
23. Mark, *The Space Station,* 189; Shapland and Rycroft, *Spacelab: Research,* 153.

24. Dennis Overbye, "Success amid the Snafus," *Discover,* November 1985, pp. 52–67.
25. Brian Barling, Shuttle Ark cartoon, 1983(?).
26. Headlines gradually shifted from J. Michael Kennedy, "Astronauts Conduct Experiments, Face New Problems," *Los Angeles Times,* May 1, 1985, p. B5; to Thomas O'Toole, "Shuttle Crew Conquering Problems in Spacelab," *Washington Post,* May 4, 1985, p. A2; to "Shuttle Returns with Data Trove," *Chicago Tribune,* May 7, 1985, p. 1, to Richard D. Lyons, "Pleased Scientists Hail Achievement as Shuttle Lands," *New York Times,* May 7, 1985, p. A1; and Jonathan Eberhart, "Spacelab: Success amid Frustration," *Science News* 127 (1985): 292–93.
27. Reports of the Spacelab 2 mission found in these newspapers in July–August 1985 were consulted: *Washington Post, New York Times, Chicago Tribune, Christian Science Monitor,* and *Los Angeles Times.*
28. William J. Broad, "Countdown Proceeding Smoothly for Space Shuttle Blastoff Today," *New York Times,* July 12, 1985, p. B5.
29. Conclusions based on survey of the *Washington Post, New York Times, Chicago Tribune, Christian Science Monitor,* and *Los Angeles Times* for two to three months surrounding each mission, as well as feature articles noted separately.
30. "Screwdriver Fixes Glitch on Challenger," *Los Angeles Times,* December 2, 1983, p. A2; Thomas O'Toole, "Pliers-Wielding Shuttle Astronauts Repair Spacelab Equipment," *Washington Post,* December 3, 1983, p. A9.
31. The press noticed Dr. Robert Parker's annoyance with workload and interruptions during the Spacelab 1 mission; see John Noble Wilford, "Shuttle Mission May Be Extended for a Day," *New York Times,* December 2, 1983, p. B24.
32. Philip J. Hilts, "Spacelab Beaming Back Discoveries," *Washington Post,* December 1, 1983, p. A1; Philip J. Hilts, "Spacelab Tests Are Yielding Breakthroughs," *Washington Post,* December 4, 1983, p. A1; Peter N. Spotts, "Earthbound Scientists Jubilant as Spacelab Data Start Pouring In," *Christian Science Monitor,* December 6, 1983, p. 6; Lee Dembart and Lee Dye, "Spacelab Called a Huge Success," *Los Angeles Times,* December 8, 1983, p. B1; Editorial, "The Newest Frontier," *Los Angeles Times,* December 9, 1983, p. F6; "Space Authorities Exult over Columbia Mission," *Washington Post,* December 10, 1983, p. A3; Peter N. Spotts, "Scientists Delight in Ocean of Data from Spacelab Mission," *Christian Science Monitor,* December 12, 1983, p. 3.
33. Arlen J. Large, "NASA Push for a Permanent Space Station Aided by European Lab Orbited by Shuttle," *Wall Street Journal,* November 29, 1983, p. 12; J. Michael Kennedy, "Shuttle Tests Look to Space Stations of Tomorrow," *Los Angeles Times,* May 5, 1985, p. A4; Richard D. Lyons, "Astronauts Stow Gear as they Head for Coast Landing: Mission Termed Success," *New York Times,* May 6, 1985, p. A1. Also, a general statement about paving the way to a space station appears in Richard D. Lyons, "Pleased Scientists Hail Achievement as Shuttle Lands," *New York Times,* May 7, 1985, p. A1.
34. Editorial, "Shuttling Toward a Space Policy," *Christian Science Monitor,* September 12, 1983, p. 24; Robert C. Cowen, "Shuttle Nears Goal: Smooth, Regular Operation," *Christian Science Monitor,* September 15, 1983, p. 15.
35. Mitchell Waldrop, "Spacelab: Science on the Shuttle," *Science* 222 (1983): 405–9, and a similar assessment after the early Spacelab missions in Cass Peterson, "Refocusing the Shuttle Toward Science," *Washington Post,* September 26, 1988, p. A4.

36. Philip H. Abelson, "Spacelab 1," *Science* 225 (1984): 127.
37. Overbye, "Success amid the Snafus," 52–67.
38. Overbye, "Success amid the Snafus," 64.
39. Valerie Neal, Tracy McMahan, and Dave Dooling, *Science in Orbit: The Shuttle-Spacelab Experience, 1981–1986* (Huntsville, AL: NASA Marshall Space Flight Center, 1988). In 1986 the Life Sciences Division at NASA Headquarters created a compendium: Thora W. Halstead and Patricia A. DuFour, eds., *Biological and Medical Experiments on the Space Shuttle 1981–1985* (Washington, DC: NASA Life Sciences Division, 1986), which included summaries of each experiment and citations for early publications of results.
40. Shapland and Rycroft, *Spacelab: Research,* especially 117–80 (quote on 153).
41. J. A. Van Allen, no title, *New York Times,* April 1, 1986, p. A31.
42. Spacelab Life Sciences-1 (STS-40, 1991), Spacelab Life Sciences-2 (STS-58, 1993), Spacelab-Mir Mission (STS-71, 1995), Life and Microgravity Spacelab Mission (STS-78, 1996), and Neurolab (STS-90, 1998). Astronaut Rhea Seddon recounts the two Spacelab Life Sciences missions in her memoir, *Go for Orbit* (Murfreesboro, TN: Your Space Press, 2015).
43. Press kits for these missions, available at www.nasa.gov, contain experiment descriptions. In addition, NASA published a series of booklets featuring the investigations planned for each mission: *Spacelab Life Sciences-1,* NP-120 (August 1989), *Spacelab Life Sciences-2* (no date), Spacelab-Mir (no date), *The Life and Microgravity Spacelab Mission* (no date), and *Neurolab* (no date).
44. Conclusions based on survey of the *Washington Post, New York Times, Chicago Tribune, Christian Science Monitor,* and *Los Angeles Times* for two to three months surrounding each mission, as well as feature articles noted separately. Kathy Sawyer of the *Washington Post* reported well on the body's response to spaceflight in "Research on Weightlessness Holds Key to Mars Missions," July 6, 1987, p. A1; "Shuttling into Orbit to Explore How the Body Adjusts," May 20, 1991, p. A3; and "Columbia Returns Safely, Filled with Scientific Data," June 15, 1991, p. A2. See also Warren E. Leary, "Shuttle's Next Mission: Better Health in Space," *New York Times,* May 21, 1991, p. C8.
45. Jay C. Buckey, Jr., and Jerry L. Homick, eds., *The Neurolab Spacelab Mission: Neuroscience Research in Space,* NASA SP 2003–535 (Houston: National Aeronautics and Space Administration, 2003). News reports included Steve Ditlea, "For Final Shuttle Science Mission, a Payload of Brain Studies," *New York Times,* April 14, 1998, p. F2, and "Columbia Soars into Space with a Menagerie on Board," *New York Times,* April 18, 1998, p. A9.
46. The primarily materials science missions included two International Microgravity Labs (STS-42 and STS-65) in 1992 and 1994; a U.S. Microgravity Lab (STS-50) in 1992; a series of four U.S. Microgravity Payload missions (STS-52, STS-62, STS-75, and STS-87) in 1992, 1994, 1996, and 1997; and two Materials Science Labs (STS-84 and STS-94) in 1997.
47. *First International Microgravity Laboratory, Second International Microgravity Laboratory, The First United States Microgravity Laboratory,* and *NASA's Microgravity Science Laboratory: Illuminating the Future,* all undated NASA publications.
48. Robert J. Naumann and Harvey W. Fleming, *Materials Processing in Space: Early Experiments,* NASA SP-443 (Washington, DC: National Aeronautics and Space Administration, 1980); Space Studies Board, National Research Council, *Toward a Microgravity Research Strategy* (Washington, DC: National Academy Press, 1992); Space Studies Board,

National Research Council, *Microgravity Research Opportunities for the 1990s* (Washington, DC: National Academy Press, 1995).

49. Warren E. Leary reporting on the first International Microgravity Laboratory mission, "International Crew Set for Shuttle Launching," *New York Times,* January 21, 1992, p. C2.

50. Two missions dedicated to astronomy were flown: Astro 1 (STS-35, 1990) and Astro 2 (STS-67, 1995). Three Atmospheric Lab in Space missions flew: ATLAS 1 (STS-45, 1992), ATLAS 2 (STS-56, 1993), and ATLAS 3 (STS-66, 1994).

51. *Astro: Exploring the Invisible Universe of Ultraviolet and X-Ray Astronomy,* NP-121, and *Astro-2: Continuing Exploration of the Invisible Universe; ATLAS: Atmospheric Laboratory for Applications and Science* and *ATLAS 1: Encountering Planet Earth,* all undated NASA publications.

52. *Columbia Accident Investigation Board Report,* vol. 1 (Washington, DC: U.S. Government Printing Office, 2003), 27.

53. James A. Van Allen, "Is Human Spaceflight Now Obsolete?," *Science* 304 (2004): 822; Matthew B. Koss, "How Science Brought Down the Shuttle," *New York Times,* June 29, 2003, p. WK13; Andrew Lawler, "Research Drought Looms After Neurolab Mission," *Science* 280 (1998): 515–16; Simon Ostrach, "Spacelab's Worth," *Science* 280 (1998): 2027, 2029.

54. Kathy Sawyer, "Glenn, Back in Space? Idea Still in NASA Orbit," *Washington Post,* January 14, 1998, p. A1; Warren E. Leary, "Glenn Kept Talking to NASA, and After 2 Years the Agency Said Yes," *New York Times,* January 17, 1998, p. A8; Francis X. Clines, "John Glenn to Go Back into Orbit, at Age 77," *New York Times,* January 18, 1998, p. A1, and "Glenn Is Ready to Say Goodbye to All That," *New York Times,* June 13, 1998, p. A7; Jeffrey Kluger, "The Right Stuff, 36 Years Later," *TIME,* January 26, 1998; Mireya Navarro, "Glenn Back on Ground, Head Still Above Clouds," *New York Times,* November 8, 1998, p. 1.

55. Lawrence K. Altman, "Studying Aging in Space? Send an Aging Astronaut," *New York Times,* January 27, 1998, p. F1; Bill Carter, "Glenn's Show, and Cronkite's," *New York Times,* July 29, 1998, p. E5; "Glenn's Mission" (cover) and Jeffrey Kluger and Dick Thompson, "John Glenn: Back to the Future," *TIME,* August 17, 1998; John Noble Wilford, "For Glenn and the Nation, a Trip Back in Time," *New York Times,* October 25, 1998, p. 1, and "Glenn Draws Huge Crowd to View His Latest Flight," October 29, 1998, p. A14; Kathy Sawyer, "The Glory Days, Light-Years Away," *Washington Post,* October 28, 1998, p. A1; "Space Aging," *New York Times,* October 19, 1998, p. A30; Jeffrey Kluger et al., "Victory Lap," *TIME,* November 9, 1998; Space Foundation, Douglas S. Morrow Public Outreach Award, 1999.

56. Kathy Sawyer, "Effect of Weightlessness on Lungs Surprising, Shuttle Study Finds," *Washington Post,* June 8, 1991, p. A2; "Astronauts Surprised by Reaction to Space," *New York Times,* June 11, 1991, p. C11; Kathy Sawyer, "Shuttle Experiments Yield Surprises on Body's Adaptation to Space Flight," *Washington Post,* September 20, 1991, p. A3.

57. Buckey and Homick, *The Neurolab Spacelab Mission,* published five years after STS-90; J. P. Downey, *Life and Microgravity Spacelab (LMS) Final Report,* NASA/CP-1998-206960 (Huntsville, AL: NASA Marshall Space Flight Center, February 1998); Fred Leslie, *Final Science Results Spacelab J* (Washington, DC: National Aeronautics and Space Administration, February 1995).

58. Neal, McMahan, and Dooling, *Science in Orbit;* Marsha Torr, *The Spacelab Scientific Missions: A Comprehensive Bibliography of Scientific Publications,* TM 108487 (Huntsville, AL: NASA Marshall Space Flight Center, April 1995), available at ntrs.nasa.gov. Robert Naumann et al., for Marshall Space Flight Center, *Spacelab Science Results Study* (Hanover, MD: NASA Center for AeroSpace Information, 2009), available at ntrs.nasa.gov. In September 2000, NASA published on compact disc the proceedings of the Spacelab Accomplishments Forum held in Washington, D.C., in March 1999 to review the results from almost two decades of research on the shuttle. Two editions of Arnauld E. Nicogossian, Carolyn Leach Huntoon, and Sam L. Pool, *Space Physiology and Medicine,* appeared in quick succession after shuttle and Spacelab results became available (Philadelphia: Lea & Febiger, 1989; repr. 1994). Space Studies Board, National Research Council, *A Strategy for Research in Space Biology and Medicine in the New Century* (Washington, DC: National Academy Press, 1998).
59. On the relevance to longer duration spaceflight, see Tamara Jernigan, "Spacelab—An Astronaut's View," *Washington Post,* July 31, 1991, p. A20; "Columbia Returns Safely, Filled with Scientific Data," *Washington Post,* June 15, 1991, p. A2. Books derived from life science research in space include Arnauld E. Nicogossian and Oleg G. Gazenko, eds., *Space Biology and Medicine,* vol. 3: *Humans in Spaceflight* (Reston, VA: American Institute of Aeronautics and Astronautics, 1993); Susanne E. Churchill, ed., *Fundamentals of Space Life Sciences,* vol. 2 (Malabar, FL: Krieger, 1997); Charles F. Sawin et al., eds., *Extended Duration Orbiter Medical Project Final Report, 1989–1993,* NASA SP-1999-534 (Houston, TX: NASA Johnson Space Center, 1999); and Jay C. Buckey, Jr., *Space Physiology* (Oxford: Oxford University Press, 2006).
60. John Noble Wilford, "Scientists Debate Missions in Space: Result of Crystal Experiments on Shuttle Doesn't Justify High Cost, Some Say," *New York Times,* November 26, 1992, p. B18; Barry L. Stoddard et al., "Mir for the Crystallographers' Money," *Nature* 360 (1992): 293–94.
61. "The Flying Ph.D.s," *Newsweek,* October 7, 1996.
62. Kathy Sawyer, "Columbia Crew Shares TV Limelight," *Washington Post,* July 6, 1992, p. A17, reporting on the first U.S. Microgravity Lab mission (STS-50).
63. Spacelab Module 1, Accession number A19990001000, on display at the Smithsonian National Air and Space Museum's Steven F. Udvar-Hazy Center since 2003.

CHAPTER 5. SPACE STATION

1. John Noble Wilford, "The Shuttle's Future: NASA Looks Toward Space Station," *New York Times,* March 16, 1982, p. C1; Philip E. Culbertson, NASA Associate Deputy Administrator, letter to John Noble Wilford, *New York Times,* March 24, 1982, with attached document, "A Space Station for America," Folder 1019, NASA Historical Reference Collection. Beggs also used "the next logical step" phrase in "Why America Needs a Space Station," a speech given in Detroit on June 23, 1982 (according to Howard E. McCurdy, *The Space Station Decision: Incremental Politics and Technological Choice* [Baltimore: Johns Hopkins University Press, 1990; repr. Washington, DC: Smithsonian Institution Press, 1997], 4).

2. Roger D. Launius, *Space Stations: Base Camps to the Stars* (Washington, DC: Smithsonian Books, 2003); *Long Range Plan of the National Aeronautics and Space Administration* (Washington, DC: National Aeronautics and Space Administration, December 16, 1959); Roland W. Newkirk, Ivan D. Ertel, and Courtney G. Brooks, *Skylab: A Chronology,* SP-4011 (Washington, DC: National Aeronautics and Space Administration, 1977), xv, 11; Arnold S. Levine, *Managing NASA in the Apollo Era,* SP-4102 (Washington, DC: National Aeronautics and Space Administration, 1982), 240–41.
3. Space Task Group Report, *The Post-Apollo Space Program: Directions for the Future* (Washington, DC: Executive Office of the President, September 15, 1969), 14.
4. *Space Station: Key to the Future,* EP-75 (Washington, DC: National Aeronautics and Space Administration, 1970).
5. Hans Mark, *The Space Station: A Personal Journey* (Durham, NC: Duke University Press, 1987); McCurdy, *The Space Station Decision.*
6. W. D. Kay, "Democracy and Super Technologies: The Politics of the Space Shuttle and Space Station *Freedom,*" *Science, Technology, and Human Values* 19/2 (1994): 131–51. See Chapter 1 notes for citations of Erving Goffman, William A. Gamson, Robert D. Benford, and David A. Snow on how organizations frame their goals and seek support.
7. James M. Beggs, "The Wilbur and Orville Wright Memorial Lecture to the Royal Aeronautical Society," London, England, December 14, 1984, Folder 009098, Space Station Brochures, NASA Historical Reference Collection; NASA Office of Program Planning and Evaluation, *Long Range Plan of the National Aeronautics and Space Administration* (December 16, 1959), and Space Task Group, *The Post-Apollo Space Program: Directions for the Future* (September 1969).
8. Beggs, "Wright Memorial Lecture."
9. John Noble Wilford, "Space and the American Vision," *New York Times Sunday Magazine,* April 5, 1981, pp. 14ff.; McCurdy, *The Space Station Decision,* 147; Kay, "Democracy and Super Technologies," 144.
10. Kay, "Democracy and Super Technologies," 139–40, 143.
11. Beggs, "A Space Station for America," Folder 1019, NASA Historical Reference Collection.
12. Hans Mark, "The Space Station—Mankind's Permanent Presence in Space," The Aerospace Medical Association Louis H. Bauer Lecture, May 17, 1984, Folder 51110, Hans Mark Speeches, NASA Historical Reference Collection. This lecture and his book, *Space Station: A Personal Journey,* are primary accounts for this analysis of the campaign of inspiration and persuasion. A strategic planning advisory committee led by former NASA administrator James C. Fletcher was named in September 1981 (see Appendix 5 in Mark, *Space Station,* 241–42), and also about the same time a Space Station Task Force on technical requirements and possible configurations was created, led by Philip E. Culbertson and John D. Hodge, both mentioned in Mark, *Space Station,* 15–16.
13. Space Station Task Force, *Space Station Program Description Document,* NASA Technical Memorandum TM-86652, March 1984, Folder 009409, Space Station Program Description Document, NASA Historical Reference Collection.
14. Various versions of "A Space Station for America," John D. Hodge Biographical File 001019 and File 009098, Space Station Brochures, NASA Historical Reference Collection.

15. Robert Freitag Biographical File 000722, NASA Historical Reference Collection; Terence Finn Biographical File 17084, NASA Historical Reference Collection; Robert Freitag, "After Apollo—the Space Station, Shuttle, and Tug," *Science Journal,* August 1970, pp. 31–38.
16. Terence Finn, various memos to John Hodge in June–July 1982, Folder 17084, Later Think Pieces by Finn, NASA Historical Reference Collection.
17. Terence T. Finn, oral history interview by Sylvia D. Fries, June 12, 1985, pp. 11–12, Folder 00659, Finn, Terence T., NASA Historical Reference Collection.
18. Mark's own account in the Bauer Lecture and *Space Station;* R. Jeffrey Smith, "Squabbling over the Space Policy," *Science* 217 (1982): 331–33.
19. Comparison of the two texts, Appendix 7 and Appendix 8 in Mark, *Space Station,* 248–50; Howell Raines, "Reagan Affirms Support for U.S. Space Program," *New York Times,* July 5, 1982, p. 8; Richard Kolulak and Carol Oppenheim, "Columbia's Landing Opens New Era in Space," *Chicago Tribune,* July 5, 1982, p. 3.
20. John Noble Wilford, "Space and the American Vision," *New York Times Sunday Magazine,* April 5, 1981, pp. 14ff.; Wilford, "Space Program Faces Hurdles Despite Cheers for Columbia," *New York Times,* April 19, 1981, p. E1; Wilford, "Will NASA's Pet Project Fizzle or Fly?," *New York Times,* October 2, 1983, p. E6; Wilford, "U.S.-Soviet Space Race: 'Hare and Tortoise,'" *New York Times,* December 22, 1983, p. A15.
21. "Space Station in the Ballot Box," *New York Times,* December 18, 1983, p. E18; cartoon by Paul Corio for the *Washington Post,* and Alex Roland, "Cost in Space," *Washington Post,* May 22, 1994, p. C1.
22. "Space Station," National Security Decision Directive 5–83, April 11, 1983, Document III-39, in John M. Logsdon, ed., *Exploring the Unknown: Selected Documents in the History of the U.S. Civil Space Program,* vol. 1: *Organizing for Exploration,* SP 4407 (Washington, DC: National Aeronautics and Space Administration, 1995), 593–95; "National Space Policy," National Security Decision Directive Number 42, July 4, 1982, Document III-38, ibid., 590–94.
23. "Revised Talking Points for the Space Station Presentation to the President and the Cabinet Council," November 30, 1983, with attached "Presentation on Space Station," December 1, 1983, Document III-40, ibid., 595–600; Finn oral history interview, June 12, 1985, pp. 13–15.
24. Ronald Reagan, "Address Before a Joint Session of the Congress on the State of the Union," January 25, 1984, available at www.reagan.utexas.edu/archives/speeches/1984/12584e.htm; also "State of the Union Message, January 25, 1984," *Public Papers of the Presidents of the United States: Ronald Reagan, 1984* (Washington, DC: Office of the Federal Register, National Archives and Records Administration, 1986): 87–95.
25. Mitch Waldrop, "The Selling of the Space Station," *Science* 223 (1984): 793–94; "An Expensive Yawn in Space," *New York Times,* January 29, 1984, p. E18; Nicholas Wade, "Stranded on the Sea of Tranquility," *New York Times,* July 20, 1984, p. A26; "The Ultimate Junket," *New York Times,* November 9, 1984, p. A30; "The Wrong Stuff," *New York Times,* November 21, 1984, p. A20.
26. James M. Beggs, *Space Station: The Next Logical Step* (Washington, DC: National Aeronautics and Space Administration, 1984?); also published in *Aerospace America,* September 1984, pp. 46ff.

27. *U.S. Congress, House Committee on Appropriations, Subcommittee on HUD-Independent Agencies, Department of Housing and Urban Development—Independent Agencies Appropriations for 1985, Part 6, National Aeronautics and Space Administration, March 27* (Washington, DC: U.S. Government Printing Office, 1984), 8.
28. John D. Hodge, "The Space Station Program Plan," *Aerospace America,* September 1984, pp. 56–59.
29. "Space Station," NASA Facts, KSC Release No. 16–86, January 1986; "Space Station," NASA Information Summaries, PMS-008 (Hqs), December 1986; all in Folder 009098, Space Station Brochures, NASA Historical Reference Collection.
30. Walter Froehlich, *Space Station: The Next Logical Step,* NASA EP-213 (Washington, DC: National Aeronautics and Space Administration, 1984), quote on p. 44.
31. "The Challenge Beyond Challenger," *New York Times,* January 31, 1986, p. A30; "Where Is NASA Headed?," *New York Times,* September 29, 1988, p. A26.
32. Andrew J. Stofan, *Space Station: The Next Logical Step* (Washington, DC: National Aeronautics and Space Administration, 1986).
33. Marcia S. Smith, "NASA's Space Station Program: Evolution of Its Rationale and Expected Uses," *Testimony before the Senate Subcommittee on Science and Space* (April 20, 2005) and "NASA's Space Station Program: Evolution and Current Status," *Testimony before the House Science Committee* (Washington, DC: *Congressional Research Service,* April 4, 2001); Launius, *Space Stations,* 111–41.
34. See the *New York Times* editorials "To Mars, Via Moscow," December 24, 1987, p. A18, and "Where Is NASA Headed?," September 19, 1988, p. A26.
35. William F. Chana, "A Suggested Name for the U.S. 'Space Station,'" *Astronautics & Aeronautics,* September 1983, p. 8, Folder 17549, NASA Names—Space Station 1981–, NASA Historical Reference Collection.
36. Dixon Otto, publisher of *Countdown,* letter to James M. Beggs, September 11, 1984, and other correspondence, Folder 17549, NASA Names—Space Station 1981–, NASA Historical Reference Collection.
37. Meeting notes recorded by Dr. Sylvia Fries to document the 1988 naming process, Folder 17549, NASA Names—Space Station 1981–, NASA Historical Reference Collection.
38. "President May Announce Name for the Space Station," *Defense Daily,* June 22, 1988, p. 292; "Reagan Picks Name for Space Station," *Houston Post,* July 29, 1988; both clippings in Folder 17549, NASA Names—Space Station 1981–, NASA Historical Reference Collection.
39. Colin L. Powell, memorandum to the President, "Naming the Space Station," July 6, 1988, and The White House Office of the Press Secretary, "Statement by Marlin Fitzwater, Assistant to the President for Press Relations," July 18, 1988, both in Folder 17549, NASA Names—Space Station 1981–, NASA Historical Reference Collection.
40. See justification of the name *Freedom* drafted in NASA Public Affairs Office, Folder 17549, NASA Names—Space Station 1981–, NASA Historical Reference Collection.
41. Associate Administrator for Space Station and Associate Administrator for Communications, memo to all Officials . . . , "Style Guidelines for 'Space Station Freedom,'" August 4, 1988, Folder 17549, NASA Names—Space Station 1981–, NASA Historical Reference Collection; "Logo Selected for Space Station Freedom Program," NASA News Release 89–68,

May 5, 1989, Folder 004554, Space Station Emblem/Patch, NASA Historical Reference Collection.

42. Franklin D. Martin and Terence T. Finn, *Space Station: Leadership for the Future,* PAM509/8–87 (Washington, DC: National Aeronautics and Space Administration, 1987); Andrew J. Stofan and Terence T. Finn, *Space Station: A Step into the Future,* PAM510/11–87 (Washington, DC: National Aeronautics and Space Administration, 1987); Leonard David, *Space Station Freedom: A Foothold on the Future,* NP-107 10/88 (Washington, DC: National Aeronautics and Space Administration, 1988); *Space Station Freedom: Gateway to the Future,* NP-137 (Washington, DC: National Aeronautics and Space Administration, 1992); "Space Station Freedom—Keystone to Human Exploration [lithograph]," NASA HqL-312.
43. "Space Station Themes," Office of Space Station, NASA Headquarters, April 9, 1986 (a set of repeated use briefing charts), Folder 009428, Space Station: Still the Next Logical Step, NASA Historical Reference Collection.
44. Boeing, *America Needs Space Station Freedom,* 1987; Boeing ads in *New York Times,* June 17, 1986, p. A17, and *Aviation Week & Space Technology,* September 15, 1986; Boeing, McDonnell Douglas, and Rockwell Public Relations, *The Space Station,* undated; McDonnell Douglas, *Space Station: Why a Manned Space Station Now?,* 1984; Martin Marietta, *Space Station,* April 1985. All in Folder 009098, Space Station Brochures, NASA Historical Reference Collection.
45. David, *Space Station Freedom: A Foothold.*
46. *Space Station Freedom: Gateway to the Future.*
47. Marcia S. Smith, "NASA's Space Station Program: Evolution and Current Status," April 4, 2001, Folder 17977, Space Station History, and Folder 17089, ISS Congressional Hearings 2001, NASA Historical Reference Collection; David Kolosta, "Space Station Plans" cartoon for the *Houston Post,* May 2, 1988, Folder 31211, Cartoons 1988, NASA Historical Reference Collection.
48. "The Case for Space Station Freedom: A Statement of Purpose," 1989, Folder 009098, Space Station Brochures, NASA Historical Reference Collection.
49. Editorials in the *New York Times:* "Adrift in Space," January 7, 1986, p. A20; "How to Regain Face in Space," May 28, 1986, p. A22; "The Poverty of NASA's Dream," February 12, 1987; "To Mars, via Moscow," December 23, 1987, p. A18; "A Rival for the Space Palace," January 15, 1988, p. A30; "The Next President's Choices in Space," October 8, 1988, p. A26; "To the Moon—and Back," July 16, 1989, p. E22; "NASA's Black Hole in Space," March 29, 1990, p. A22; "Space Yes; Space Station No," June 6, 1991, p. A24; "NASA's Untouchable Folly," July 14, 1991; "The Wrong Space Station," July 29, 1992, p. A20.
50. NASA Information Summaries PMS-008B (Hqs), "Space Station *Freedom,*" March 1992; "The Wrong Space Station," *New York Times,* July 29, 1992, p. A20. See also a similar dismantling of the case for a space station in "NASA's Untouchable Folly," *New York Times,* July 14, 1991, p. E18.
51. *Space Station Freedom Media Handbook,* May 1992, ii–7. First edition published in April 1989.
52. Smith, "NASA's Space Station Program: Evolution and Current Status," *Testimony before the House Science Committee* (Washington, DC: Congressional Research Service, April 4, 2001); Jeff Margulies, "Sam's Trailer Park," for *Houston Post* (1987) in *Aviation Week & Space Technology,* December 24, 1990, p. 49; Gary Brookins, "*@#! Budget Cuts," *Rich-*

mond Times Dispatch, 1993; Chip Bok, "Space Station Wagon," for *Akron Beacon Journal,* in *Washington Post,* June 26, 1993, p. A25; "Space Station Design No. 407," *Houston Chronicle,* June 9, 1993; and Dana Summers, "What Do You Mean, Model?" *Orlando Sentinel,* 1992; Chuck Asay, "Space Station Freedom," *Colorado Springs Gazette Telegraph,* 1992.

53. John H. Gibbons, Assistant to the President for Science and Technology, letter to Administrator Daniel Goldin, February 18, 1993, and Space Station Talking Points, Questions and Answers sent from the White House, Folder 12680, Space Station Redesign, 1993, NASA Historical Reference Collection.
54. NASA Press Conference, March 11, 1993, transcript, Folder 17041, Space Station-General, 1993, NASA Historical Reference Collection; Kathy Sawyer, "Space Station Redesign Due by June 1," *Washington Post,* March 11, 1993, p. A11; Warren E. Leary, "Under White House Pressure, NASA Is Designing a Cheaper Space Station," *New York Times,* March 12, 1993, p. A16; "Gibbons Outlines Space Station Redesign Guidance," NASA Press Release 93–64, April 6, 1993, and "Space Station Redesign Advisory Members Named," NASA News Release 93–59, April 1, 1993, both available at mira.hq.nasa.gov/history; "Statement of the President," June 17, 1993, Press Releases April–June 1993, Record 33764, The White House, Office of the Press Secretary.
55. Sean Holton, "A Lost Decade Deflates Space Station Dreams," *Orlando Sentinel,* May 30, 1993, p. A-16.
56. "Station Redesign Team to Submit Final Report," NASA News Release 93–104, June 4, 1993; "Statement by Daniel S. Goldin on Space Station Redesign," NASA News Release 93–104, June 17, 1993, Press Releases April–June 1993, Record 33764, both available at mira.hq.nasa.gob/history. Bryan D. O'Connor et al., *Space Station Redesign Team Final Report to the Advisory Committee on the Redesign of the Space Station* (Washington, DC: National Aeronautics and Space Administration, June 1993); Kathy Sawyer, "Administration Warned on Space Station Redesign," *Washington Post,* March 16, 1993, p. A10; Kathy Sawyer, "Space Panel Chair Predicts Freedom Alternates Won't Fly," *Washington Post,* May 21, 1993, p. A4; "Panel Leader Backs Space Station Design," *New York Times,* May 21, 1993; Kathy Sawyer, "Space Station Redesign Team Scrambles to Meet Basic Goals," *Washington Post,* May 10, 1993, p. A1; "Station Redesign Team Evaluates Its Options," NASA Headquarters Bulletin, May 17, 1993; "NASA Rethinks the Space Station," *Science* 260 (1993): 1228–29; "Station Redesign Team to Submit Final Report," NASA News Release 93–104, June 4, 1993; Bryan O'Connor, "Space Station Redesign Team Final Presentation to the Advisory Committee for the Redesign of the Space Station [briefing charts]," June 7, 1993, in Folder 16281, Space Station Redesign Team Final Presentation, NASA Historical Reference Collection.
57. "NASA Pares Field in Effort to Rechristen Space Station," *New York Times,* November 22, 1993, p. A8. An inconclusive process lasted from a late 1993 call for suggestions to the first crew's arrival on the station in 2000. Bryan D. O'Connor, NASA Director of Space Station Transition, memo to Distribution, "Renaming of the Space Station Program," September 24, 1993; NASA Early Bird News Summary, "Space Station Name," November 22, 1993; "A Space Station in Search of a Name," MSNBC, October 29, 2000; William Harwood, "Crew of 'Alpha' Males Moves into Space Station," *Washington Post,* November 3, 2000, p. A2; Steven Siceloff, "NASA Yields to Alpha Name," *Florida Today,*

February 1, 2001, p. A1, all in Folder 17549, NASA Names—Space Station 1981–, NASA Historical Reference Collection; The White House, Office of the Press Secretary, "Statement of the President," with "Fact Sheet: Space Station Redesign," June 17, 1993, and John H. Gibbons, Assistant to the President for Science and Technology, letter to NASA Administrator Daniel Goldin [guidance for an implementation plan], June 24, 1993, all in Folder 12680, WH/Advisory Committee on Redesign of the Space Station, NASA Historical Reference Collection; "Statement by Daniel S. Goldin, NASA Administrator, on Space Station Redesign," NASA News Release, June 17, 1993, Press Releases April–June 1993, Record 33764, available at mira.hq.nasa.gov/history.

58. "Jettison the Space Station," *New York Times,* June 23, 1993, p. A22; Nick Anderson, "Black Hole," *Washington Post,* December 7, 1998, after first publication in the *Louisville Courier-Journal* editorial page (December 1, 1998); NASA pamphlets, *Six Reasons Why America Needs the Space Station* (no date, but probably late 1993 or early 1994 because it uses the term International Space Station instead of *Freedom*) and "Exploding the Myth: Facts About the Space Station," International Space Station Fact Book, June 10, 1994.

59. Habitation Module graphics and descriptions, File 9620, Station Habitation Module, NASA Historical Reference Collection; also in same file, "Note to Editors," NASA News Release 87–110, September 15, 1987, and "Users to View Space Station Modules," NASA News Release 87–141, September 24, 1987; Thomas H. Maugh II, "Cooped Up," *Los Angeles Times,* September 28, 1987, p. Q11; Michael and Courcy Hinds, "An Efficiency Apartment for 8 Weightless People," *New York Times,* June 30, 1988, pp. C1, C6; Ray Spangenburg and Diane Moser, "At Home on the Space Station," *Final Frontier,* August 1988, 23–26.

News articles on space station downsizing and the habitation module include William Broad, "NASA Reduces Cost and Role of Its Orbiting Space Station," *New York Times,* March 5, 1991, p. A1; Kathy Sawyer, "NASA Unveils Scaled-Back Space Station," *Washington Post,* March 22, 1991, p. A11; Phil Patton, "Lost in Space: Living Room for the Crew," *New York Times,* September 8, 1994, pp. C1, C6; Florence Williams, "Putting a Room of One's Own in Orbit," *New York Times,* December 30, 1999, p. F1.

60. Tom Toles, "Last Piece Almost in Place," *Washington Post,* December 17, 2006, p. B6, File 44235, Cartoons 2006, NASA Historical Reference Collection.

61. John Noble Wilford's articles on this topic in the *New York Times* included "Space Station: Costs and Allies' Role Raise Doubts," February 8, 1987, pp. 1, 32; "Shades of Sputnik: Who's Ahead in Space?," January 3, 1988, p. E7; "Giving New Purpose to the Space Program," December 11, 1990, p. C14.

62. Editorials in *New York Times:* "Opening the Space Station," November 1, 2000, p. A34; "A Space Station out of Control," November 25, 2001, p. WK10; "Is the Space Station Necessary?," August 14, 2005, p. C11; "Back to the Moon, Permanently," December 9, 2006, p. A18.

63. Peter Carlson, "Is NASA Necessary?," *Washington Post Magazine,* May 30, 1993, pp. 11ff.

64. NASA and the Office of Science and Technology Policy, "Launching a New Era in Space Exploration," February 1, 2010, available at www.whitehouse.govfiles/documents/ostp//press_release_files/Joint%20Statement%202.2.pdf; National Space Policy of the United States, June 28, 2010, available at history.nasa.gov/national_space_policy_6–2.pdf; NASA

Authorization Act of 2010, Pub. Law No. 111–267 (October 11, 2010), available at www.congress.gov/111/plaws/publ267/PLAW-111publ267.pdf.

65. Launius, *Space Stations*, 134; John M. Logsdon, "The Space Station Is Finally Real," *Space Times: Magazine of the American Astronautical Society* 34 (1995): 23.

CHAPTER 6. PLANS

1. NASA Office of Program Planning and Evaluation, *Long Range Plan of the National Aeronautics and Space Administration* (Washington, DC: National Aeronautics and Space Administration, December 16, 1959).
2. The President's Scientific Advisory Committee, chaired by Donald Hornig, "Report of the Ad Hoc Panel on Man-in-Space," December 16, 1960.
3. Jerome Wiesner, "Report to the President-Elect of the Ad Hoc Committee on Space," January 10, 1961.
4. Dwayne A. Day, "The Von Braun Paradigm," *Space Times*, November–December 1994, pp. 12–15; "Paradigm Lost," *Space Policy* 11 (1995): 153–59.
5. Space Task Group Report, *The Post-Apollo Space Program: Directions for the Future* (Washington, DC: Executive Office of the President, September 15, 1969), i.
6. "Summary Report, Future Programs Task Group" (Washington, DC: National Aeronautics and Space Administration, 1965).
7. Donald P. Hearth, *Outlook for Space: Report to the NASA Administrator*, SP-386 (Washington, DC: National Aeronautics and Space Administration, 1976).
8. Phillip Blackerby, "History of Strategic Planning," available at blackerbyassoc.com/history.html; Government Performance and Results Act (GPRA), Pub. L. 103–62, 103rd Cong., 1st Sess. (1993).
9. Blackerby, "History of Strategic Planning"; Carter McNamara, "All About Strategic Planning," Free Management Library, available at managementhelp.org/strategicplanning.
10. Warren G. Bennis and Burt Nanus, *Leaders: Strategies for Taking Charge* (New York: Harper & Row, 1997).
11. Mark Rhodes, "Strategic Thinking: Mental Frames and Strategy," available at strategybydesign.org/mental-frames.
12. National Aeronautics and Space Administration Authorization Act, FY 1985, Title II, Sections 203–204, Pub. Law No. 98–361 (July 16, 1984); Robert G. Torricelli (Democrat from New Jersey) of the House Science and Technology space science and applications subcommittee, quoted in Paul Mann, "Washington Overview," *Commercial Space*, Spring 1986, p. 13.
13. John F. Murphy, NASA Assistant Administrator for Legislative Affairs, letter to Senator Pete V. Domenici, Chairman, Committee on the Budget, January 25, 1984; Assistant Administrator for Legislative Affairs, memo to the Administrator, "National Space Commission," June 1, 1984, both in NASA Historical Reference Collection; "National Commission on Space," Exec. Order 12490 (October 12, 1984).
14. Report of the National Commission on Space, *Pioneering the Space Frontier: An Exciting Vision of Our Next Fifty Years in Space* (New York: Bantam Books, May 1986); members are listed on pp. 194–95; Amy B. Zegart, "Blue Ribbons, Black Boxes: Toward a Better

Understanding of Presidential Commissions," *Presidential Studies Quarterly* 34 (2004): 366–93.

15. Marcia S. Smith, National Commission on Space Executive Director, correspondence with author, September 30, 2012.
16. Sally K. Ride, *Leadership and America's Future in Space: A Report to the Administrator* (Washington, DC: National Aeronautics and Space Administration, August 1987).
17. Ride, *Leadership and America's Future in Space,* 54, 51.
18. Thomas O. Paine, official NASA biography, available at history.nasa.gov/Biographies/paine.html.
19. Ride, *Leadership and America's Future in Space,* 6.
20. Nicholas Wade, "Stranded on the Sea of Tranquility," July 20, 1984, p. A26; "Adrift in Space," January 7, 1986, p. A20; "How to Regain Face in Space," May 28, 1986, p. A22; "The Poverty of NASA's Dreams," February 12, 1987, p. A30, all in *New York Times.*
21. John Noble Wilford, "Exploration of Mars Is Advised as Goal for NASA," March 18, 1987, p. B6, and "The Allure of Mars Grows as U.S. Searches for New National Goal," March 24, 1987, pp. C1, C3, both in *New York Times;* "Settlements on Moon and Mars Proposed by Panel," *New York Times,* March 25, 1986, p. C8; James Fisher, "40-Year Goal: Manned Martian Base," *Washington Post,* March 25, 1986, p. A5; Chris Reidy, "Space Travel Blueprint Targets Mars," *Orlando Sentinel,* July 23, 1986; Philip M. Boffey, "Panel Urges Lunar Station as Step to Mars," *New York Times,* August 18, 1987, p. C3; Kathy Sawyer, "Study Urges Missions to Moon, Mars," *Washington Post,* August 18, 1986, p. A1; Laurie McGinley, "Ride Urges NASA to Examine Earth and Explore Mars," *Wall Street Journal,* August 18, 1987.
22. Philip M. Boffey, "NASA Faces Year of Crucial Decisions," January 28, 1987, *New York Times,* p. D27; Larry D. Spence, "NASA's Pie in the Sky," *New York Times,* October 14, 1987, p. A35; Kathy Sawyer, "White House Accused of Failing to Give NASA Sense of Mission," *Washington Post,* August 17, 1987, p. A3.
23. Michael Isikoff, "NASA Chief Hails Push for Mars Colony: Others Fear Public Will Laugh at Idea," May 24, 1986, p. A3, and Kathy Sawyer, "White House Accused of Failing to Give NASA Sense of Mission," August 17, 1987, p. A3, both in *Washington Post;* "NASA Studying Major Space Initiatives," NASA News Release: 87–37, March 17, 1987, Folder 009452, NASA Historical Reference Collection.
24. Brad Bumsted, "Will Sally Ride's Space Report Fly?," *Gannett News Service,* August 7, 1987; Mark Carreau and William Clayton, "Space Experts Ask Reagan, Congress to Act on Report," *Houston Chronicle,* August 18, 1987; "America's Future in Space," *Washington Post,* August 20, 1987, p. A22; "Space Leadership: A Direction Emerges," *Aviation Week & Space Technology,* August 24, 1987; "Getting NASA Back on Track," *TIME,* August 31, 1987; all in Folder 001794, Ride Report, NASA Historical Reference Collection.
25. James C. Fletcher, NASA Administrator, letter to Representative Robert A. Roe, Chairman, Committee on Science, Space and Technology, and also to Senator Ernest F. Hollings, Chairman, Committee on Commerce, Science and Transportation, February 18, 1988, and NASA News Note to Editors: Review of National Space Commission Report, February 19, 1988, all in Folder 12775, National Commission on Space, NASA Historical Reference Collection.

26. Michael Specter, "Astronaut Sally Ride to Leave NASA," May 27, 1987, p. A4; "Astronaut Sally Ride Ends 10-Year Career at NASA," September 27, 1987, p. A12; both in *Washington Post;* William J. Broad, "Ride Quitting Astronauts for Stanford," *New York Times,* May 27, 1987, p. B7. The following year, NASA's Office of Exploration published another report on the topic, *Beyond Earth's Boundaries: Human Exploration of the Solar System in the 21st Century* (Washington, DC: National Aeronautics and Space Administration, 1988).
27. "Remarks on the 20th Anniversary of the Apollo 11 Moon Landing," 1989-07-20, in *Public Papers of the President: George Bush, 1989,* available at bushlibrary.tamu.edu/research/public_papers.php?id=712&year=1989&month=all.
28. Kathy Sawyer, "Bush Urges Commitment to New Space Exploration; Objectives Include Moon Base, Mars Mission," *Washington Post,* July 21, 1989, p. A4.
29. *Report of the 90-Day Study on the Human Exploration of the Moon and Mars,* November 1989, available at history.nasa.gov/90_day_study.pdf. For an account of the tensions and frustrations behind this report, considered a failure by the White House, see Thor Hogan, *Mars Wars: The Rise and Fall of the Space Exploration Initiative,* SP-2007-4410 (Washington, DC: National Aeronautics and Space Administration, 2007).
30. *Report of the 90-Day Study,* "Executive Summary," 1–1; "Preface," 2–2.
31. Neil Armstrong in a pre-launch press conference on July 5, 1969, quoted in James R. Hansen, *First Man: The Life of Neil A. Armstrong* (New York: Simon & Schuster, 2005), 399, citing Norman Mailer, *Of a Fire on the Moon* (New York: Little, Brown and Co., 1969), 43–44.
32. National Research Council, Committee on Human Exploration of Space, *Human Exploration of Space* (Washington, DC: National Academy Press, 1990), ix; *Report of the Advisory Committee on the Future of the U.S. Space Program* (Washington, DC: U.S. Government Printing Office, December 1990); Willis Hawkins, Robert Jastrow, et al., *A Report on the Lunar and Mars Initiatives* (Washington, DC: George C. Marshall Institute, 1990); *America at the Threshold: Report of the Synthesis Group on America's Space Exploration Initiative* (Washington, DC: U.S. Government Printing Office, May 1991); U.S. Congress, Office of Technology Assessment, *Exploring the Moon and Mars: Choices for the Nation*, OTA-ISC-502 (Washington, DC: U.S. Government Printing Office, July 1991); George C. Marshall Institute, *New Directions in Space: A Report on the Lunar and Mars Initiatives* (Washington, DC: George C. Marshall Institute, 1990); *Exploring the Moon and Mars: Choices for the Nation*, Office of Technology Assessment Report, OTA-ISC-502 (Washington, DC: U.S. Government Printing Office, July 1991). Thor Hogan's *Mars Wars* reviews several of these studies. George Bush, "Remarks at the Texas A&I University in Kingsville [Commencement Address]," May 11, 1990, available at bushlibrary.tamu.edu/research/public_papers.
33. National Research Council, *Human Exploration of Space,* 21.
34. Mark Albrecht, *Falling Back to Earth: A First Hand Account of the Great Space Race and the End of the Cold War* ([Lexington, KY]: New Media Books, 2011), 126–24, 137–39, gives an insider's account of the political impetus for this advisory committee.
35. *Report of the Advisory Committee on the Future of the U.S. Space Program;* see especially pages 5–6, 28–31; "How to Put Space in Its Place," *New York Times,* December 12, 1990, p. A22.
36. Albrecht, *Falling Back to Earth,* 137–39, quotation on 139; Hogan, *Mars Wars,* 125–30.

37. Synthesis Group, *America at the Threshold,* iv, 2–15; Hogan, *Mars Wars,* 125–30.
38. David S. F. Portree, *Humans to Mars: Fifty Years of Mission Planning, 1950–2000,* SP-2001-4521 (Washington, DC: National Aeronautics and Space Administration, 2001), 77; Hogan, *Mars Wars,* 2, 14, 159–66.
39. *Vision 21: The NASA Strategic Plan* (Washington, DC: National Aeronautics and Space Administration, 1992).
40. *NASA Strategic Plan* (Washington, DC: National Aeronautics and Space Administration, 1995).
41. The White House, Office of the Press Secretary, "President Bush Announces New Vision for Space Exploration Program, Remarks by the President on U.S. Space Policy at NASA Headquarters" and "President Bush Announces New Vision for Space Exploration Program," with documents "Fact Sheet" and "A Renewed Spirit of Discovery," both January 14, 2004, available at history.nasa.gov/Bush%20SEP.htm.
42. "The Call of Distant Worlds," February 9, 2003, p. WK14, and "The Challenge Ahead in Space," July 6, 2003, p. WK8, both in *New York Times.*
43. "On to Mars?," January 15, 2004, p. A20; Tom Toles cartoon, January 15, 2004, p. A20; Mike Allen and Eric Pianin, "Bush Outlines Space Agenda; President Calls for Moon Trip by 2020," January 15, 2004, p. A1; all in *Washington Post.*
44. *The Vision for Space Exploration,* NP-2004-01-334-HQ (Washington, DC: National Aeronautics and Space Administration, 2004).
45. Report of the President's Commission on Implementation of United States Space Exploration Policy, *A Journey to Inspire, Innovate, and Discover* (Washington, DC, June 2004).
46. Jennifer L. Rhatigan, ed., *Constellation Program Lessons Learned,* vol. 1: *Executive Summary,* SP-6127 (Washington, DC: National Aeronautics and Space Administration, June 2011), 4–8, 12–13.
47. The White House, Office of Science and Technology Policy, Press Release, "U.S. Announces Review of Human Spaceflight Plans," May 7, 2009, available at www.whitehouse.gov/files/documents/ostp/press_release_files/NASA%20Review.pdf; The Review of U.S. Human Space Flight Plans Committee, *Seeking a Human Spaceflight Program Worthy of a Great Nation,* October 2009, available at history.nasa.gov/AugustineCommfinal.pdf.
48. Jeff Parker, "Obama Space Plan," *Florida Today,* April 16, 2010; "NASA Turnaround," *Florida Today,* December 8, 2012.
49. Michael J. Neufeld, "The Von Braun Paradigm and NASA's Long-Term Planning for Human Spaceflight," in Steven Dick, ed., *NASA's First 50 Years: Historical Perspectives,* SP-2010–4704 (Washington, DC: National Aeronautics and Space Administration, 2010), 325–47.
50. *Human Space Exploration Framework Summary,* January 12, 2011, available at www.nasa.gov/exploration/new_space_enterprise/home/heft_summary.html.
51. In 2014–15, the NASA website was rife with images and posts on the themes of "Next Giant Leap" and "Journey to Mars." Examples include the July 14, 2014, announcement of "NASA's Next Giant Leap," a poster titled "America's Next Giant Leap," and an announcement "NASA Seeks Ideas for Where on Mars the Next Giant Leap Could Take Place," June 25, 2015; all available by keyword search at www.nasa.gov.

CHAPTER 7. MEMORY

1. Remarks available at www.nasa.gov/pdf/535146main_11_0412_Bolden_final_STS 1_ Anniv.pdf. Image available at www.nasa.gov/centers/kennedy/multimedia/images/2011-04-12.html.
2. George W. Bush, "Remarks at the National Aeronautics and Space Administration," January 14, 2004, available at www.presidency.ucsb.edu/ws/index.php?pid=72531&st=&st1= and also history.nasa.gov/Bush%20SEP.htm.
3. Jeff Parker, "One Small Step for Mars," *Florida Today,* January 12, 2004, and "Moon Mars Money," *Florida Today,* November 18, 2004, File 37486, Cartoons 2004, NASA Historical Reference Collection, also available at www.cagle.com/author/jeffparker; Tom Toles, "Nothing for Something," *Washington Post,* January 19, 2004, p. A20, File 37486, Cartoons 2004, NASA Historical Reference Collection. From the *New York Times:* David E. Sanger and Richard W. Stevenson, "Bush Backs Goal of Flight to Moon to Establish Base," January 15, 2004, p. A1; "Bush's Space Vision Thing," January 15, 2004; Kenneth Chang, "The Allure of an Outpost on the Moon," January 13, 2004; Warren E. Leary, "NASA, Money for Mars, but from Elsewhere," February 3, 2004, and "Committee on Space is Optimistic on Devising Plan to Reach Mars," February 12, 2004.
4. Steve Breen, "Uncle Sam," *San Diego Union Tribune,* 2005; Shooman, "Checklist," *Star-Ledger,* 2005; Jeff Parker, "Launch Madness, The Road to Delay," *Florida Today,* March 16, 2006, at www.politicalcartoons.com/cartoon/0ac46b7e-7cd9-4738-8d17-cff3fb6db7c2.html; Steve Benson, "Need a New Radiator," *Arizona Republic,* July 2005.
5. Daryl Cagle, "Elderly Space Shuttle," *Slate,* July 30, 2005, available at www.politicalcartoons.com/cartoon/da916a7e-1a2f-404a-9f05-721853164880.html; Mackay, "NASA Sends Shuttle Back to Space," Reuters, July 26, 2005; Dana Summers, "It's Hard to Believe He's Only 25!," *Orlando Sentinel,* 2006.
6. *NASA Transition Management Plan,* JICB-001, August 2008, available at www.nasa.gov/mission_pages/transition/home/transition_plan.html; *Space Shuttle Program Artifacts Information Pamphlet,* NASA, August 2009, available at www.nasa.gov/mission_pages/transition/home/transition_plan.html.
7. "Request for Information on Space Shuttle Orbiter Placement," NASA, January 2010, available at www.nasa.gov/mission_pages/transition/home/int_orbiter_rfi.html. Space history curators at the National Air and Space Museum discussed the fate of the orbiters several times in 2008–10 while updating their Collections Rationale and Collecting Plan. Based on their experience curating Saturn V launch vehicles permanently situated elsewhere, their consensus was that the museum should seek one flown orbiter rather than assume permanent responsibility for all of them (author's recollection).
8. NASA Office of the Inspector General, Special Report, *Review of NASA's Selection of Display Locations for Space Shuttle Orbiters,* August 25, 2011, available at oig.nasa.gov/audits/reports/FY11/Review_NASAs_Selection_Display_Locations.pdf.
9. Susan Crane, "Memory, Distortion, and History in the Museum," *History and Theory* 36 (1997): 45; Kendall R. Phillips, ed., *Framing Public Memory* (Tuscaloosa: University of Alabama Press, 2004).
10. Susan A. Crane, ed., *Museums and Memory* (Stanford, CA: Stanford University Press, 2000).

11. The Kennedy Space Center Visitor Complex in Florida, National Museum of the United Space Air Force in Ohio, and Museum of Flight in Seattle planned and built large adjacent space halls.
12. "Newspaper Front Pages (Discovery's Flyover)," April 17, 2012, National Air and Space Museum compilation.
13. As co-chair of the museum's Welcome Discovery planning team, the author participated in these and other decisions. Related planning documents are filed in the Smithsonian Institution Archives.
14. Wayne Clough, "Discovery Arrival Remarks," April 17, 2012, at Dulles International Airport, and "Discovery Welcome Remarks," April 19, 2012, at the National Air and Space Museum Steven F. Udvar-Hazy Center; "Remarks by Senator John Glenn, Smithsonian National Air and Space Museum, Welcome Discovery Ceremony, Steven F. Udvar-Hazy Center, April 19, 2012, 11:00 a.m.," April 19, 2012, Chantilly, Virginia.
15. Media and social media analytics collected by the National Air and Space Museum.
16. Roger D. Launius, "American Memory, Culture Wars, and the Challenge of Presenting Science and Technology in a National Museum," *Public Historian* 29 (2007): 13–30; Lonnie G. Bunch III, "In Museums at the National Level: Fighting the Good Fight," in James B. Gardner and Peter S. LaPaglia, *Public History: Essays from the Field* (Malabar, FL: Krieger, 1999), 345–56; Bryan Hubbard and Marouf A. Hasian, Jr., "Atomic Memories of the *Enola Gay:* Strategies of Remembrance at the National Air and Space Museum," *Rhetoric & Public Affairs* 1 (1998): 363–85; Edward T. Linenthal and Tom Engelhardt, eds., *History Wars: The Enola Gay and Other Battles for the American Past* (New York: Metropolitan Book Henry Holt, 1996).
17. The author was co–lead curator for this *Moving Beyond Earth* exhibition gallery; she wrote much of the script, selected many of the artifacts, and worked with fellow curators, designers, and educators on the look and function of the exhibition. She also wrote *Discovery: Champion of the Space Shuttle Fleet* (Minneapolis: Zenith Press, 2014).
18. Smithsonian National Air and Space Museum, *Moving Beyond Earth* exhibition, first stage opened in 2010, second stage completed in 2013; see airandspace.si.edu/exhibitions/gal113/mbe/index.cfm.
19. California Science Center Foundation Samuel Oschin Air and Space Center Project Overview, undated fact sheet.
20. Damien Cane, "Celebrating U.S. Future in Space, Hopefully: Calculated to Wow," *New York Times,* April 28, 2010, p. A14; NASA Kennedy Space Center, Space Shuttle Atlantis promotional materials, 2011–13.
21. "'Forever Remembered' Exhibit Honoring Challenger and Columbia Opens at Kennedy Space Center Visitor Complex," Kennedy Space Center, July 1, 2015, available at media.kennedyspacecenter.com/kennedy/forever-remembered-exhibit-honoring-challenger-and-columbia-opens-at-kennedy-space-center-visitor-complex.htm; "'Forever Remembered' Shares Enduring Lessons of Challenger, Columbia," NASA, June 27, 2015, available at www.nasa.gov/feature/forever-remembered-shares-enduring-lessons-of-challenger-columbia; Marcia Dunn, "Challenger, Columbia Wreckage on Public Display for 1st Time," Associated Press, August 2, 2015, available at news.yahoo.com/challenger-columbia-wreckage-public-display-1st-time-133542243.html; Carole Blair and Neil Michel, "Com-

memorating in the Theme Park Zone: Reading the Astronauts Memorial," in Thomas Rosteck, ed., *At the Intersection: Cultural Studies and Rhetorical Studies* (New York: Guilford Press, 1999), 29–83.

22. Upon landing Columbia, STS-1 mission commander John Young used this phrase, often echoed thereafter in praise of the shuttle.
23. "Wings in Orbit—Wayne Hale Interview," March 13, 2011, available at www.nasaspaceflight.com.
24. Traci Watson, "NASA Administrator Says Space Shuttle Was a Mistake," *USA Today*, September 27, 2005, p. A1; John Schwartz, "NASA Official Questions Agency's Focus on the Shuttle," *New York Times*, December 9, 2006, p. A16; Wayne Hale and Helen Lane, eds., *Wings in Orbit: Scientific and Engineering Legacies of the Space Shuttle, 1971–2010*, SP-2010-3409 (Washington, DC: U.S. Government Printing Office, 2010), 9.
25. Hale, *Wings in Orbit;* "Wings in Orbit—Wayne Hale Interview."
26. Longtime critics included scientists James A. Van Allen and Robert L. Park, historian Alex Roland, and science journalist Timothy Ferris. Van Allen and Roland are cited below. Park's articles include "Pork Barrel in Low-Earth Orbit," *New York Times*, April 18, 1993, p. E19, and "Shelving the Star Trek Myth," *New York Times*, July 12, 1997, p. 21. Ferris's most pointed piece was "Ground NASA and Start Again," *New York Times*, March 16, 1992, p. A17. See also Malcolm W. Browne, "Worthy Projects Suffer Because of the Shuttle, Critics of NASA Charge," *New York Times*, April 7, 1981, p. C4; Warren E. Leary, "Debate over the Shuttle Fleet's Value to Science Has Been Raging from the Beginning," *New York Times*, February 10, 2003, p. A21.
27. J. A. Van Allen, untitled op-ed, *New York Times*, April 1, 1986, p. A31; James A. Van Allen, "Space Science, Space Technology and the Space Station," *Scientific American* 254 (1986): 32–39; James A. Van Allen, "Is Human Spaceflight Now Obsolete?," *Science* 304 (2004): 822; Alex Roland, "The Shuttle: Triumph or Turkey?," *Discover*, November 1985, pp. 29–49; Alex Roland, "NASA's Manned-Space Nonsense," *New York Times*, October 4, 1987, p. E23; Alex Roland, "Barnstorming in Space: The Rise and Fall of the Romantic Era in Spaceflight, 1957–1986," in Radford Byerly, Jr., *Space Policy Reconsidered* (Boulder, CO: Westview Press, 1989), 33–52.
28. *The Space Shuttle: A Horizon Guide*, a Horizon/BBC Production, 2012; *Space Shuttle: Final Countdown*, a Smithsonian Networks Production, 2011; *Space Shuttle, 1981–2011: Stories from 30 Years of Exploration*, Air & Space Smithsonian Collector's Edition, Spring 2010 and 2011 editions; Piers Bizony, *The Space Shuttle: Celebrating Thirty Years of NASA's First Space Plane* (Minneapolis: Zenith Press, 2011); Dan Winters, *Last Launch: Discovery, Endeavour, Atlantis* (Austin: University of Texas Press, 2012); *The History of the Space Shuttle, Collector's Edition* (2012).
29. Roger D. Launius, John Krige, and James Craig, *Space Shuttle Legacy: How We Did It and What We Learned* (Reston, VA: American Institute of Aeronautics and Astronautics, 2013). The most comprehensive technical reference is expected to be Dennis R. Jenkins, *The Space Shuttle: The History of the National Space Transportation System*, 4th ed., to be published in 2017.
30. Roger D. Launius, "Assessing the Legacy of the Space Shuttle," *Space Policy* 22 (2006): 226–34; J. M. Logsdon, "The Space Shuttle Program—a Space Policy Failure?," *Science*

232 (1986): 1099–1105; John M. Logsdon, "The Decision to Develop the Space Shuttle," *Space Policy* 2 (1986): 103–19; Joseph N. Pelton, "The Space Shuttle—Evaluating an American Icon," *Space Policy* 26 (2010): 246–48.

31. John M. Krige, Angelina Long Callahan, and Ashok Maharaj, *NASA in the World: Fifty Years of International Collaboration in Space* (New York: Palgrave Macmillan, 2013).
32. Alex Roland, "How We Won the Moon," *New York Times Review of Books,* July 17, 1994, p. 1.
33. Linda Billings, "Wowing the Public: The Shuttle as Cultural Icon," in Launius et al., *Space Shuttle Legacy,* 299–321.

Index

Page numbers in *italics* indicate to illustrations.